VERTICAL URBANISM

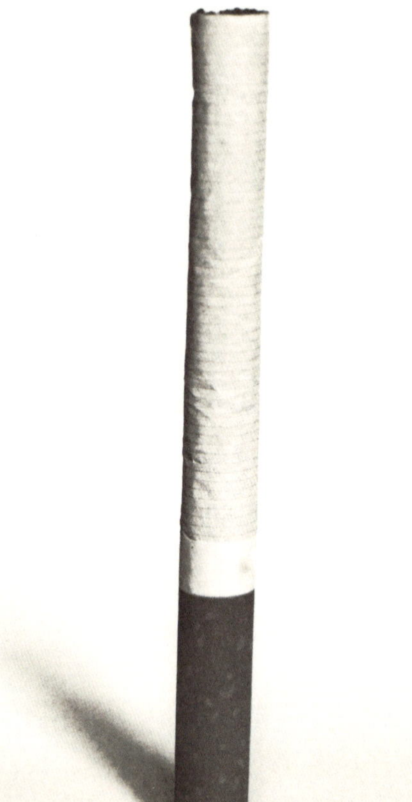

Introduction

Vertical Urbanism is a 4th Year Design Studio at Seoul National University, Department of Architecture. The Studio is run by Professor Peter W. Ferretto and looks at ways to prevent cities expanding outwards towards suburbia and beyond by offering a vertical alternative. The site for this studio was the "Dream Hub" redevelopment in Yongsan Seoul, designed by Daniel Libeskind.

Peter W. Ferretto,

Seoul, April 2012

Thank you Note

Research for this book was supported by a fund from Heerim Architects. We would like to express our thanks Mr. Jeong, CEO of Heerim Architects for his kind support and backing for this joint collaboration between Seoul National University and University of Pennsylvania. We also owe ours thanks to Ali Rahim, Professor of Architecture at UPenn for his time and dedication in making this joint program possible.

Special Thanks also to

We would also like to express our gratitude to the following people who helped in the making of this book:

Seoul National University

Professors
Jin Kyoon Kim
Du Nam Choi
Hong Gun Park
Bong Hee Jeon
Seung Hoy Kim

Guest Critics
Peter Choi

Assistant
Hwajung Song

University of Pennsylvania

Dean
Marilyn Jordan Taylor

Chairman of Department
Dr. David E. Leatherbarrow

Director of Finance
Chris Cataldo

Guest Critics
Evan Douglas
Irina Verona
Helene Furjan

9 Steps

1_Introduction

This semester's site is huge. We will be working on the Dream Hub site currently being master planned by Daniel Libeskind and designed by leading international architects. The site sits west of Yongsan station and is currently a rail depot. The semester is divided into 2 phases; phase 1 consists of group work where teams of three students will develop their own master-plan and phase two where individual students will develop a vertical typology building within their master-plan. This project will be run as a joint design studio between Seoul National University and University of Pennsylvania.

Film: Man on Wire

2_Book

Our first task will be to make a 50x50 cm book vertical urbanism reference book. Taking inspiration from O.M. Ungers "Morphologie-City Metaphors" book of 1982, we will look at concepts for vertical dwelling via images, signs, metaphors, models, analogies. Where Unger's book focused on urbanism in the flat/plan domain we will seek to explore a new form via sections and three dimensions. The book will have one format, and all students will have to work with the same layout. Research areas: Evolution of the skyscraper, vertical cities (NY/HK/Chicago/Seoul), programme, natural analogies, materials.

Period: 2 weeks

3_Site

Site analysis will be divided into three parts, each team researching a specific topic.
Topic 1 – Past/Present conditions of the site. Programme, history, geography, etc…both from micro and macro perspective. Topic 2 – Future conditions. Analysis of the Dream-Hub project and other scheme presented at the competition. Topic 3 – Planning. Regulations for the master-plan, regulations for high rise in Seoul – including maximum height, escape strategies, others…

Period: 2 weeks

4_Vertical Urbanism Master-plan

The teams will start designing their COUNTER URBAN PROPSALS to the Dream Hub, from the conceptual to outline massing. Importantly you will have to define the limits and scope to your proposal. Teams will have to address, programme (public/private), the public domain, parks, and connections with context, macro scale, circulation both vertical and horizontal and a critical position, not simply a variant in forms.

Period: 3 weeks

5_Exhibition

Coinciding with the visit of the U. Penn studio, all teams will put together and exhibition of the work to date and have reviews leading to the second phase of the project.

Period: 1 week

6_Tower(s) selection

Students will be working alone for this part of the project. From their joint master-plan developed in the first phase: student will have to select a tower or series of towers to develop. The programme will have to include 40% public institutional space such as: Park/library/cemetery/schools/hospitals/anything NOT commercial, the remaining 60%, including has to be planned by the student: private residential/commercial/hotels/offices/etc···

Period: 3 weeks

7_Architectural Development

Student will have to develop strategies for the following:
Tower Plans/Structural Grid/Floor Area/Core/Floor Area Ratio/Height/Orientation

Period: 2 weeks

8_Final Presentation

Students are to prepare:
Design report 50x50 cm
6 A1 panels portrait showing the whole project process, from urban analysis, masterplan to eventual tower proposal,
4 panels will be dedicated to the individual towers

Period: 2 weeks

9_Final Review in Philadelphia

All students will travel to U. Penn for their final joint presentation, followed by a architectural tour of Philadelphia and New York.

Period: 1 week

FREEDOM IS
FOR WHICH MANY MEN H
IN UNMARKED GRAVES WITHIN
THIS SQUARE LIE THOUSANDS
OF UNKNOWN SOLDIERS OF
WASHINGTON'S ARMY WHO DIED
OF WOUNDS AND SICKNESS DURING
THE REVOLUTIONARY WAR

IGHT
DIED IN DARKNESS
THE INDEPENDENCE AND LIBERTY
YOU POSSESS ARE THE WORK OF
JOINT COUNCILS AND JOINT
EFFORTS OF COMMON DANGERS,
SUFFERINGS AND SUCCESS.
PEOPLE
States of

Part I

Verticality

The studio started by looking at the notion of verticality, inspired by Oswald Mathias Ungers book: Morphologie: City Metaphors, we embarked on producing our own account of Vertical living. Students working in groups were asked to explore concepts, images and references that expanded on the idea. The format of the images was fixed to be 50x50 cm square images in black and white.

The following images are a compendium of the themes that were discussed which range from the obvious to the more esoteric reading.

Accumulation

[NOUN] An accumulation of something is a large number of things which have been collected together or acquired over a period of time.

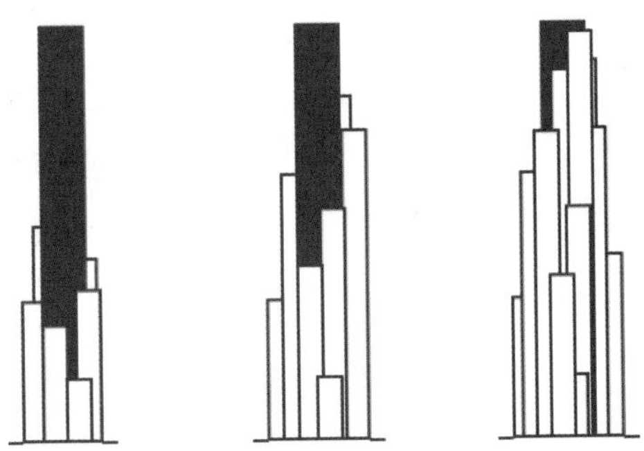

Fantasy

[NOUN] A fantasy is a pleasant situation or event that you think about and that you want to happen, especially one that is unlikely to happen.

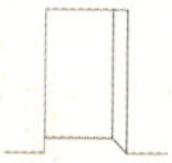

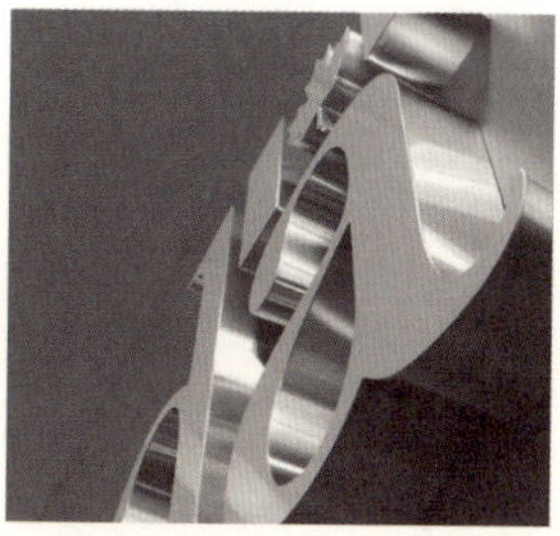

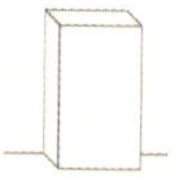

Suspend

(VERB) If something is suspended from a high place, it is hanging from that place.

Floating

(NOUN) the act of someone who floats on the water
(ADJ) having little or no attachment

Below

[PREP] If something is below something else, it is in a lower position. [PREP] If something is below a particular amount, rate, or level, it is less than that amount, rate, or level.

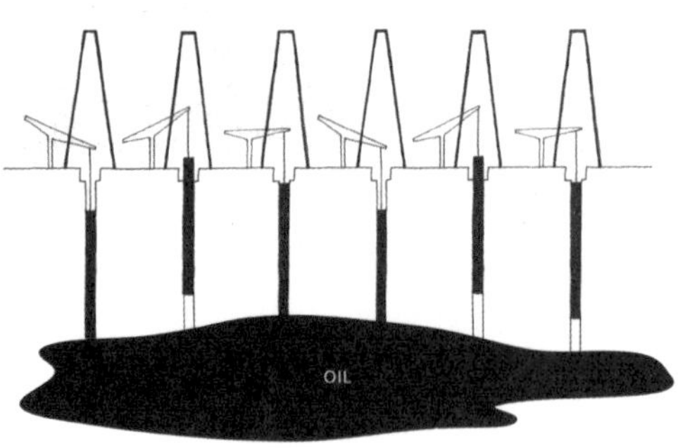

Spiral

(NOUN) A spiral is a shape which winds round and round, with each curve above or outside the previous one.

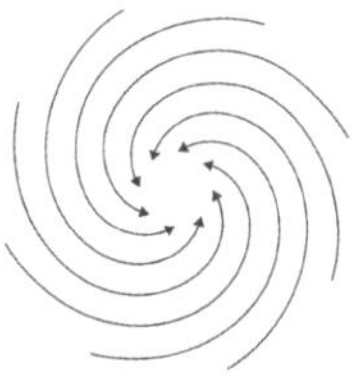

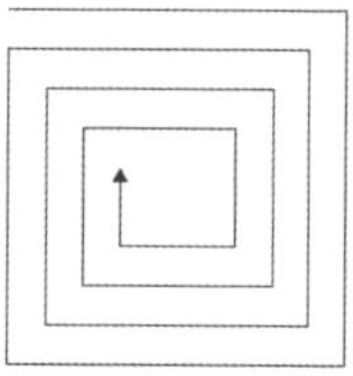

connection

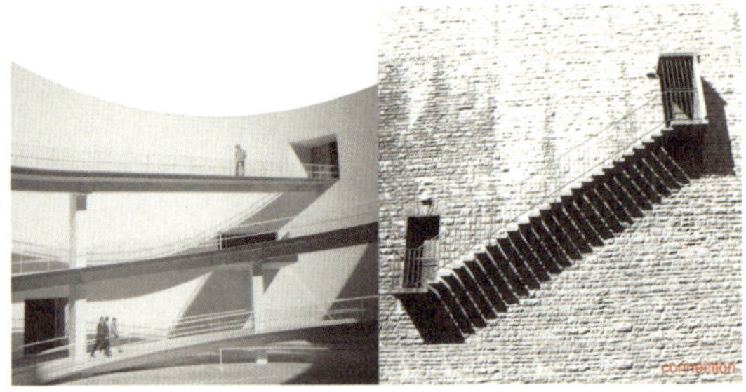

connection

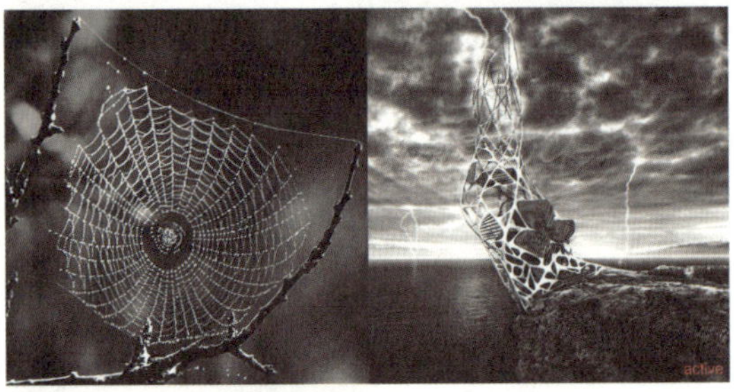

active

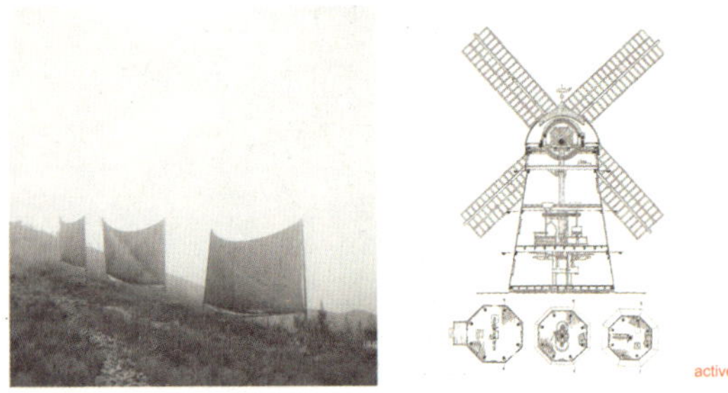

active

articulation

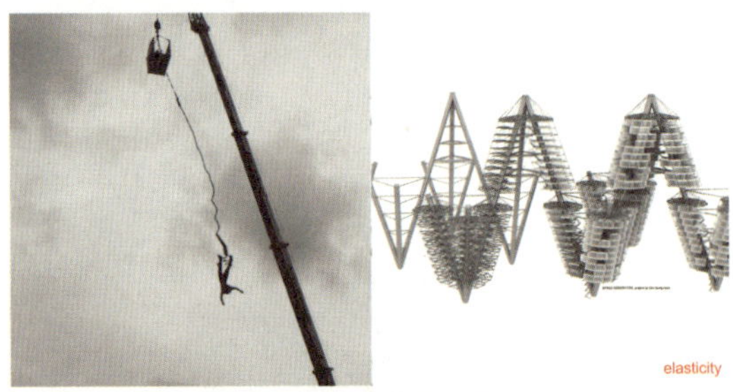

elasticity

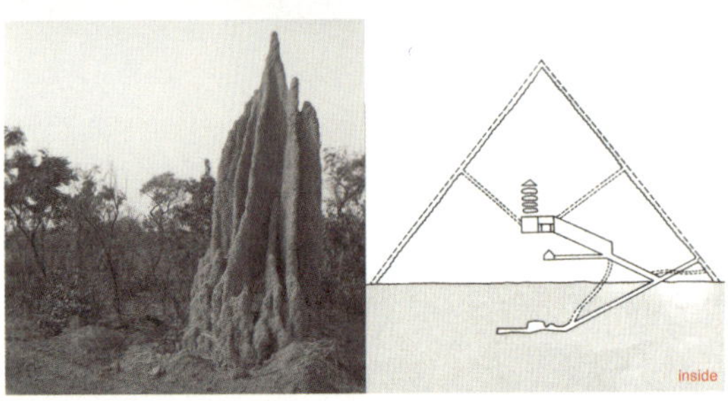

inside

in between

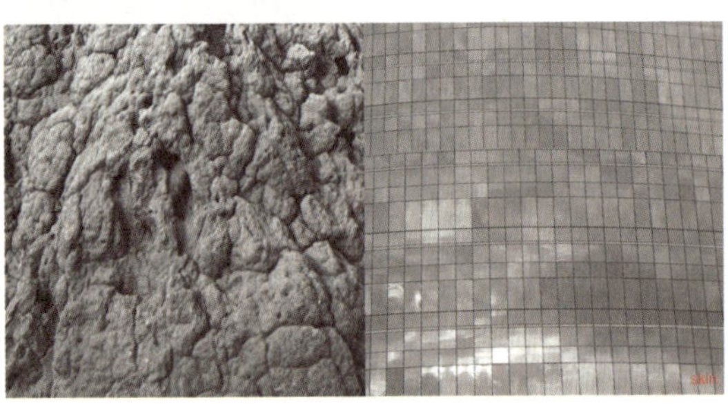

skin

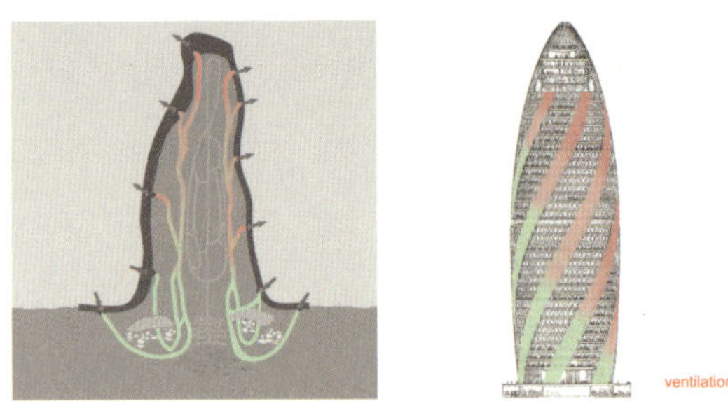

ventilation

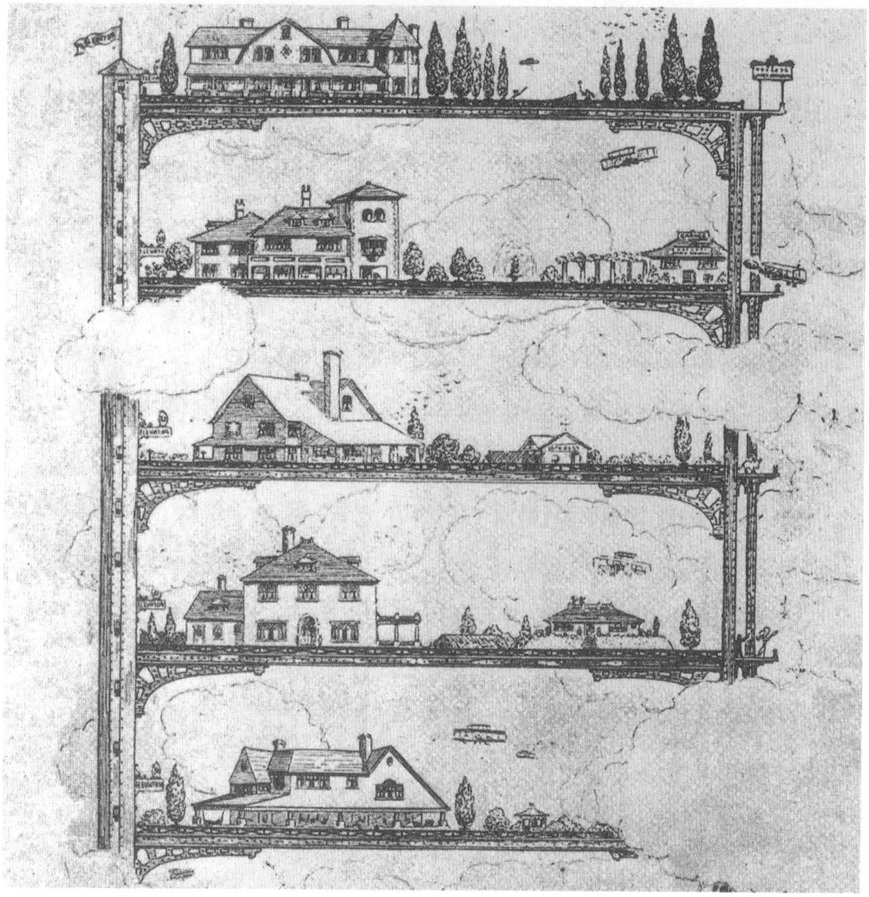

Stack

(NOUN) A stack of things is a pile of them.
(VERB) arrange in stacks

fantasy
vertigo

in between
stacking
connection
addition
subtraction
tornado(beginning)
illusion

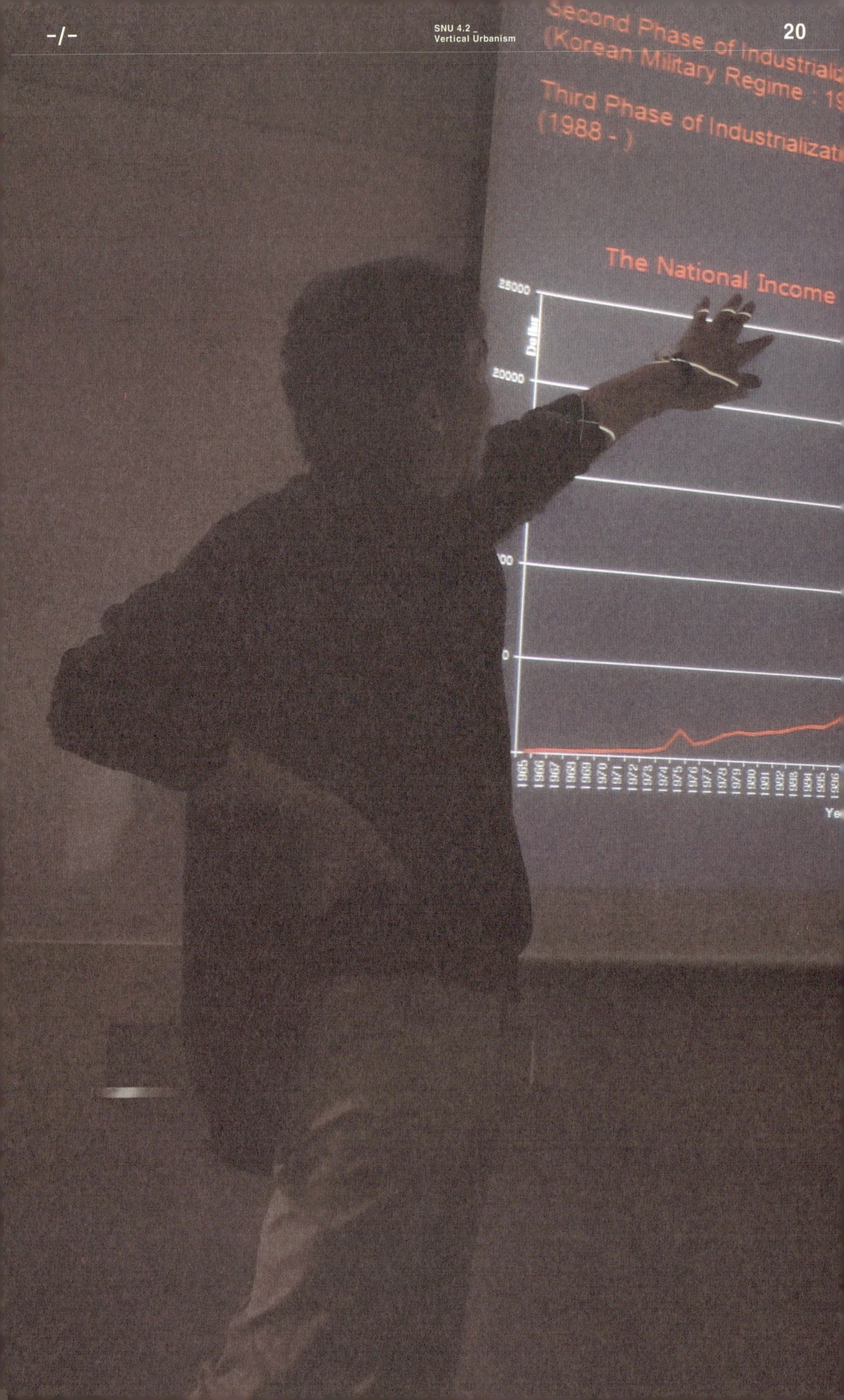
The National Income
(1988 -)
25000
20000
Dollar

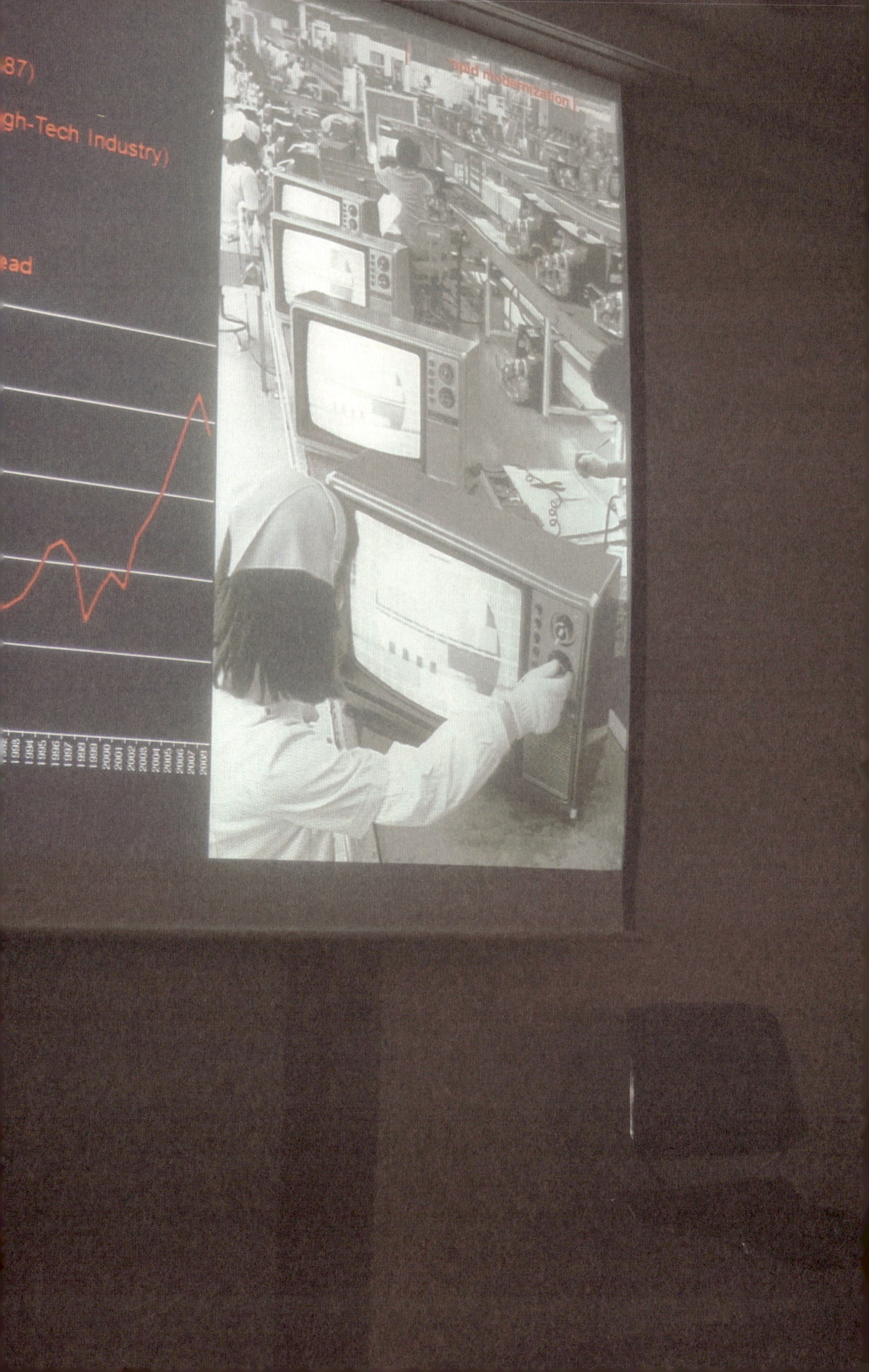
87)
gh-Tech Industry)
ead
apid modernization

Part II

Site

The Yongsang Rail Depot site is huge vacant site in the heart of the city. It's a site that originally fell outside Seoul's city walls and is strategically connected to three important infrastructural nodes: the Rail link, the US military base and the Han river. All three elements will undergo major changes in the future, lead by the US army vacating their original spot to make way to one of the biggest public parks in the world.

At the centre of the site today lies the huge I-Mall, which is both a mall and a major transport interchange, with subway, national and regional rail connections, as well as offering a comprehensive commercial and entertainment experience.

Past / Present

History

In the modern history of Korea, Yongsan worked as the entrance to the Old Seoul. In 1890s, there was a main port of travellers and stuffs in the riverside of Hangang, and it was also connected to palace district with Mancho river. So this city can be flourished with commerce and transportation.

In 1900s government made railroad and tram line, and in 1910, Yongsan was occupied by military base camp of the Japan. The empire of Japan also urbanized this city, so Yongsan still has grid system from that era. After that, In 1845 Korea could be the independent country from Japan, and 5 years later, there was the Korean War. The city was destroyed.

In 1960s, Japanese military base camp changed to the American basecamp, then prostitute and the Itaewon street was activated. On the contrary Mancho river was dimolished. In 1976, Korean government developed high density apartment in Yongsan.

In the future, US military base camp will move to the outside of Seoul, so that area will be changed to the civil park. And the district of Yongsan station will be developed as International business center of Seoul.

Physical Condition

The urban complexity of Yongsan's site makes it necessary to focus on the elemental structural conditions and illustrate them as a tool for the continuing design process. These four topics are the density of site's adjacent areas, surrounding heights and their relation to the mountains, structural and defining borderlines and pregnant landmarks in close or distant relation.

Caused by the site's size and the complex position in the network of the city, the basic research results in a, from point of the site, subjective view, which allows to see the important physical factors summarized in one diagram as subsequent assistant for the continuing design process, which therefore just can be understood as one elementary part out of many.

OKI

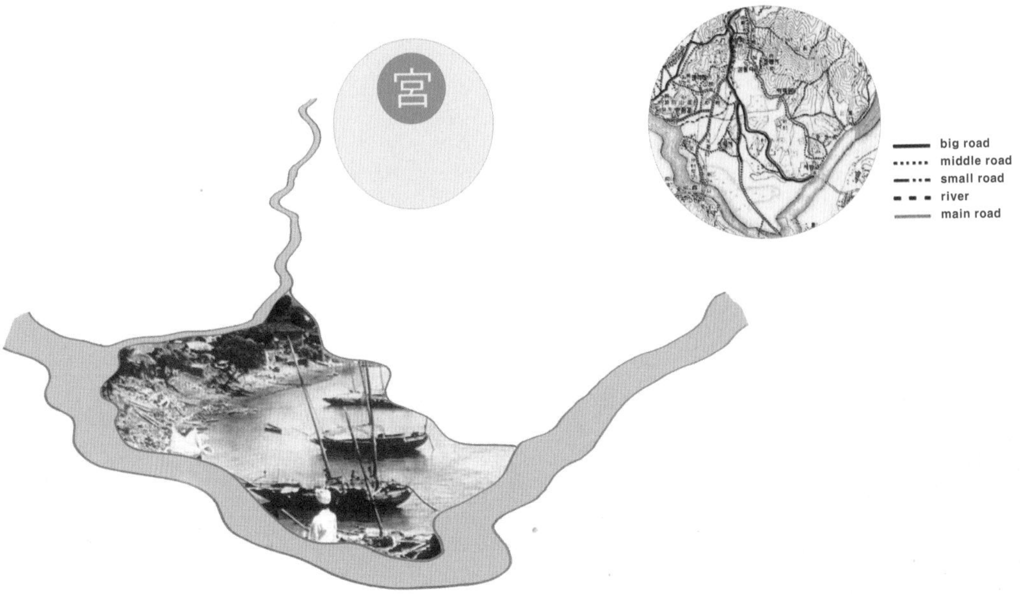

1895 Main port of traveller and stuff going to the palace district with Mancho river

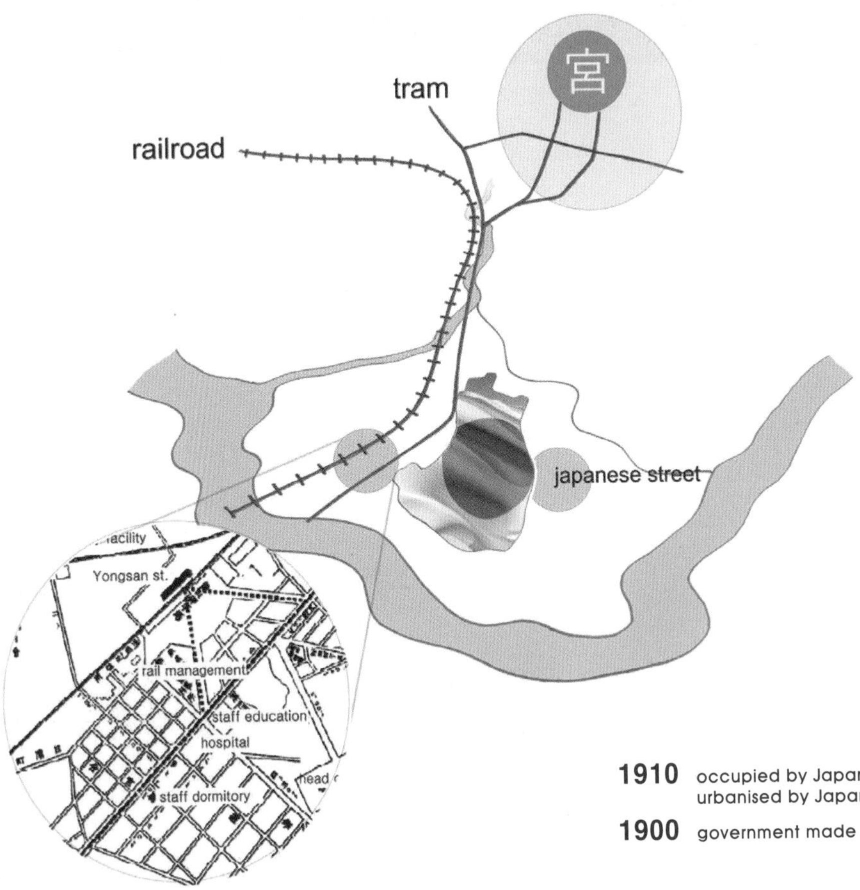

1910 occupied by Japan basecamp
urbanised by Japanese

1900 government made railroad and tram line

1945 independence from Japan colonization

1950 Korean war

1976 high density apartment development

1960 occupied by American basecamp
activation of prostitute and Itaewon street
Mancho river dimolished

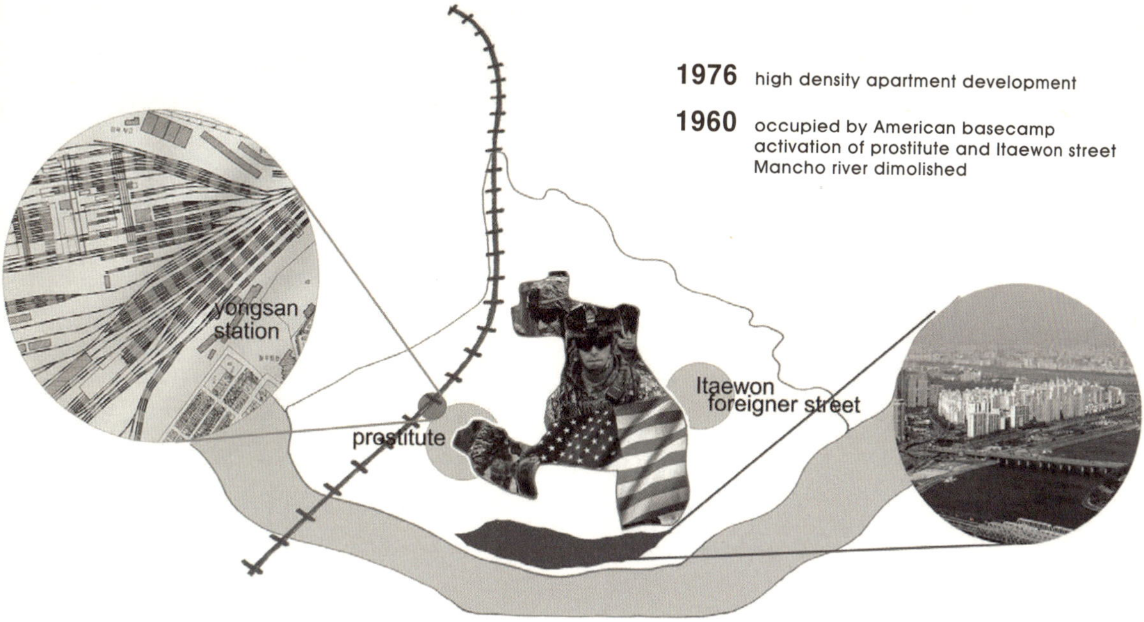

History of Yongsan

Present basecamp -> civil park
Yongsan station -> international business center

A - TRAFFIC

B- HOUSING

C - NATURE

a - car

b - apartment

c - green

a - train

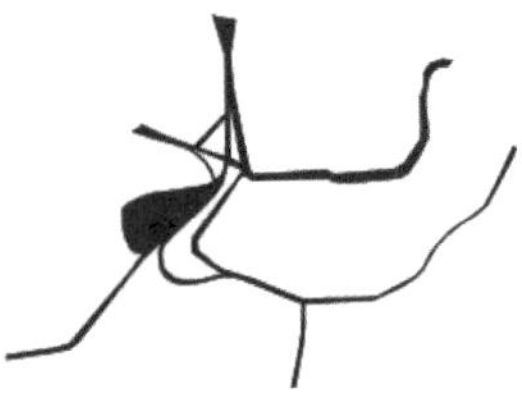

b - high apartment

c - water

People

People occupy Yongsan, and changes the area with various activities. We analyzed diverse programs in Yongsan, and made 2-dimensional index of people.

There are four topics concerned with people, Traffic, Housing, Nature, Facility. So, the visual informations of car and train in the topic of Traffic, apartments and high apartments of Housing, green area and water - also river - of Nature, entertainments, markets, accommodations, embassies, military base and schools of Facility, are collected in one map.

Various programs occupies the city and sometimes they are overlapped one another. This index of programs shows the spatial use of the people of Yongsan. It also reveals the collisions and harmonies of the people's diverse activity.

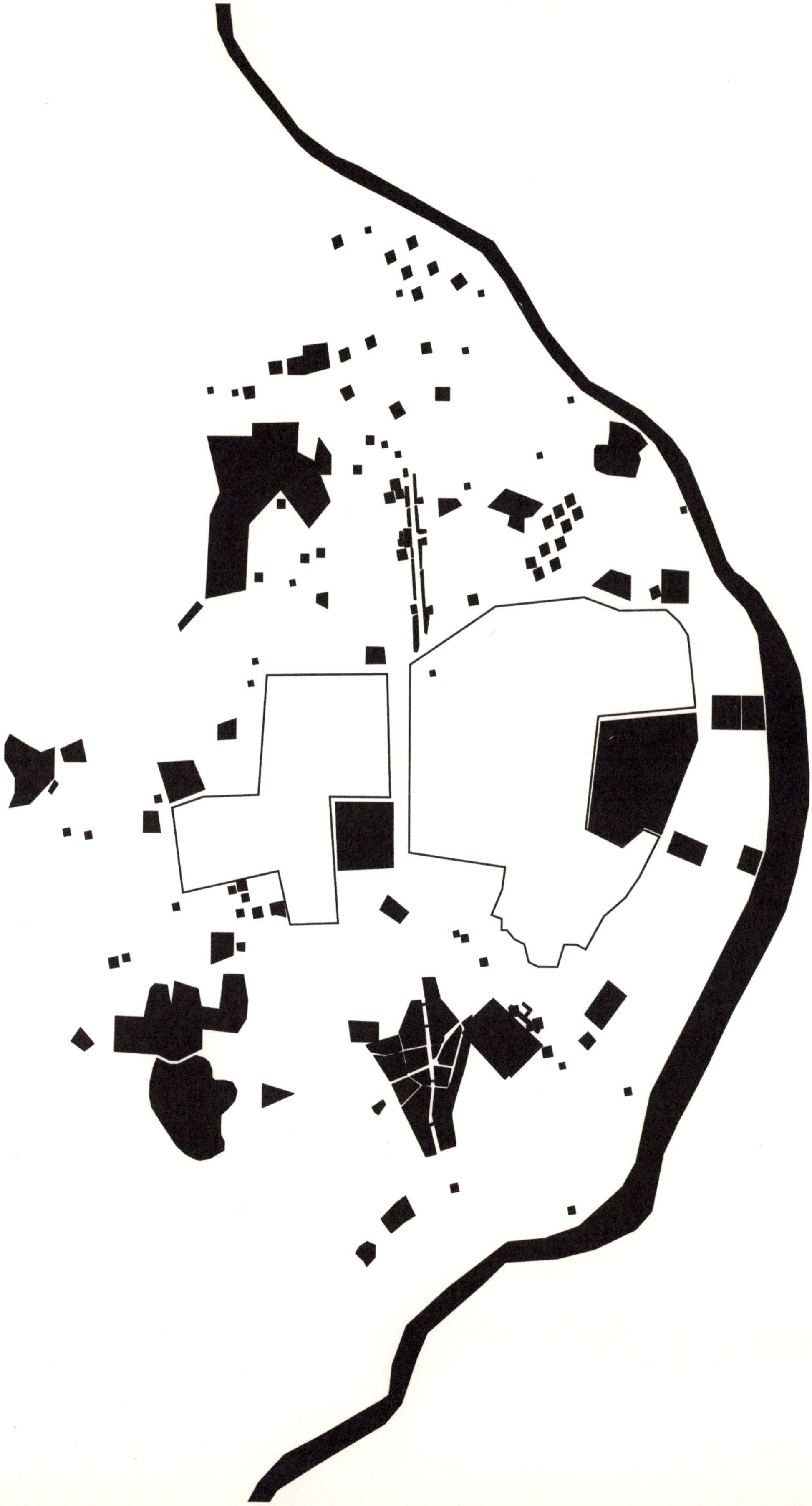

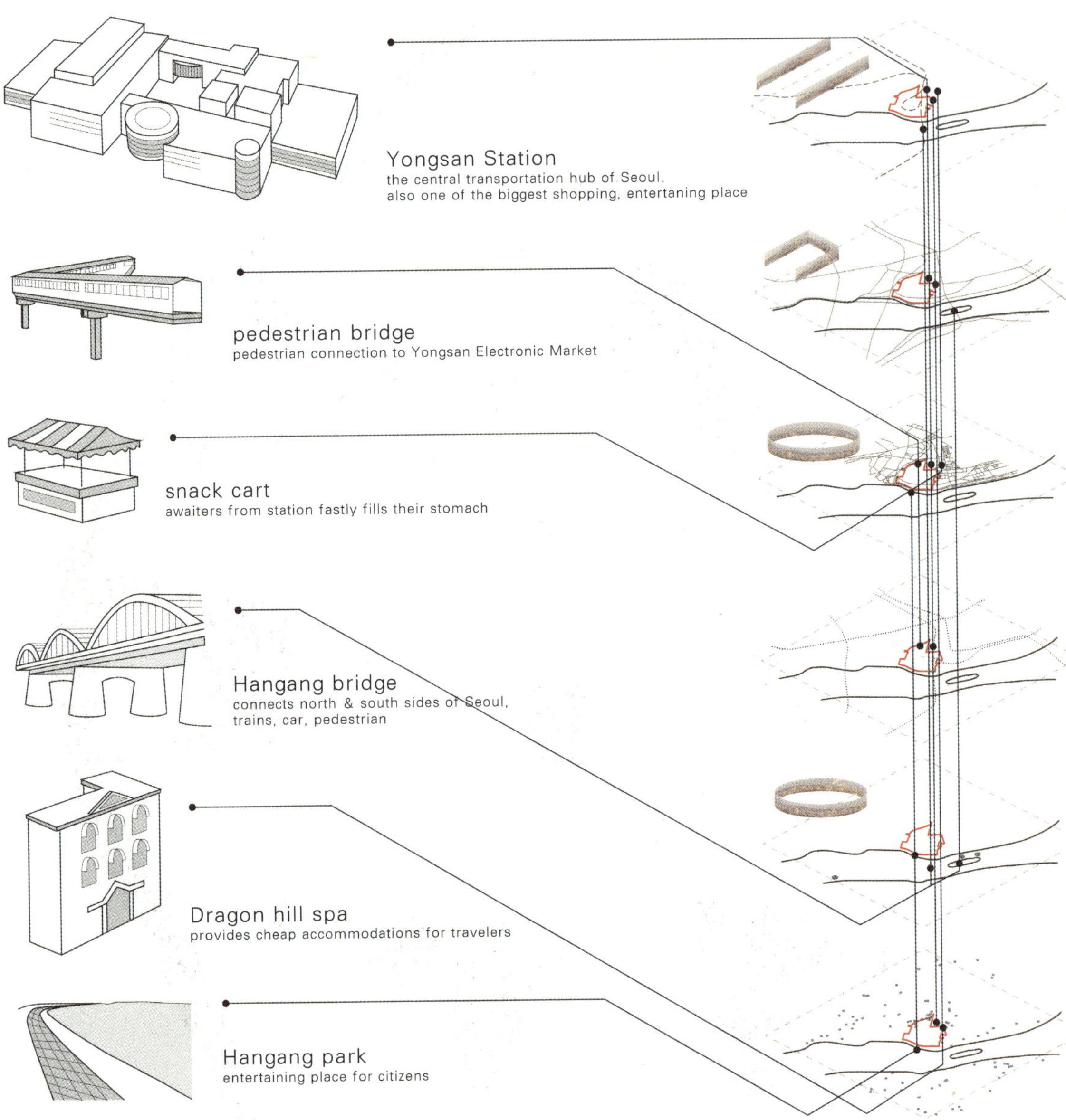
Yongsan Station
the central transportation hub of Seoul.
also one of the biggest shopping, entertaning place
pedestrian bridge
pedestrian connection to Yongsan Electronic Market
snack cart
awaiters from station fastly fills their stomach
Hangang bridge
connects north & south sides of Seoul,
trains, car, pedestrian
Dragon hill spa
provides cheap accommodations for travelers
Hangang park
entertaining place for citizens

Yongsan, the city of traffic

Yongsan is the central and hub of traffic in Seoul. Traffic is one of the important element when we analyze the Yongsan. There are six most influential traffics in Yongsan, Trains, Cars, Subways, Water vehicles, Pedestrians, Parkig lots. They make complex flows in the city. That flows change the condition of the ground and make the movements of the people.

Different traffics have the different speed and frame, so they provide people different point of view how to see the city. They give various experience to the people.

There are several spots where different traffics meet, and that spots are the important event place in Yongsan, like 'Yongsan station' and 'Dragon Hill Spa'. In Yongsan, Traffic rules the city, and when we look inside the traffics, we can easily catch the activity of the people.

railroad

Gyeongbu line KTX
Honam line KTX
Gyeonguii line Train
railroad facilities

road

9 main roads
75,941 cars

pedestrian

211,209 people

subway

5 lines
12 stations

water

3 docks
boat taxi
cruise ship

parking

6,265 parking lots
85,543 planes

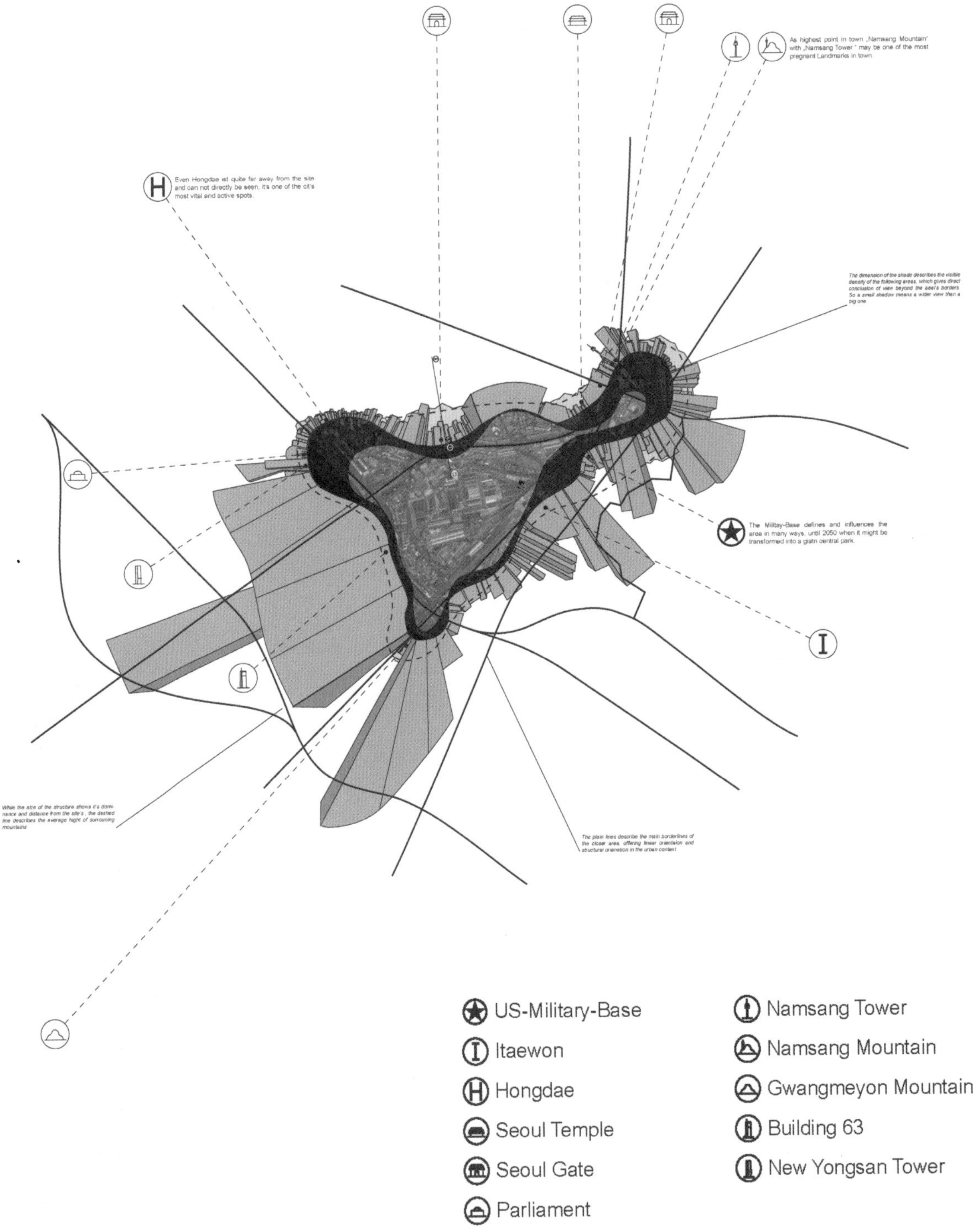
As highest point in town „Namsang Mountain“ with „Namsang Tower “ may be one of the most pregnant Landmarks in town
Even Hongdae ist quite far away from the site and can not directly be seen, it's one of the cit's most vital and active spots.
The dimension of the shade describes the visible density of the following areas, which gives direct conclusion of view beyond the seal's borders. So a small shadow means a wider view than a big one
The Militay-Base defines and influences the area in many ways, until 2050 when it might be transformed into a giatn central park.
While the size of the structure shows it's dominance and distance from the site's , the dashed line describes the average hight of surrouning mountains
The plain lines describe the main borderlines of the closer area, offering linear orientaion and structural orienation in the urban context
US-Military-Base
Itaewon
Hongdae
Seoul Temple
Seoul Gate
Parliament
Namsang Tower
Namsang Mountain
Gwangmeyon Mountain
Building 63
New Yongsan Tower

Seoul

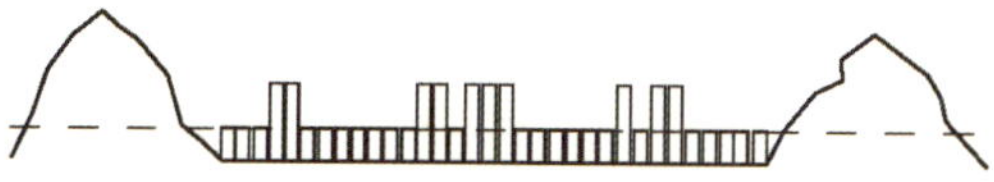

The urban complexity of Yongsan's site makes it necessary to focus on the elemental structural conditions and illustrate them as a tool for the continuing design process. These four topics are the density of site's adjacent areas, surrounding heights and their relation to the mountains, structural and defining borderlines and pregnant landmarks in close or distant relation.

Caused by the site's size and the comlex position in the network of the city, the basic research results in a, from point of the site, subjective view, which allows to see the important physical factors sumarised in one diagram as subsequent assistant for the coninuing design process, which therefore just can be understood as one elemtar part out of many.

Yongsan - Site in Seoul

The site is surrounded by a variation of great high and lower, but even great volumes, generating various different views and conditions in every direction, even changing by the individual position inside the planning-area.

As a strong vertical oriented city with lots of vertical volumes, Seoul also is a wide spreading city with very regular hights across it's area, where the greater amount of lower buildings (1-5 floors) are interupted by, usually uniform and lined up, vertical towers, visible all over the city as massive volume with fine structurized surfaces.

While the city itself seems very chaotic and organized without any higher plan or concept, built structures abrupt end at the beginning of rising mountains, due which seem to be islands in the liquid spreading city. So their effect as hight defining elements is even strengthenend by the green antural contrast to the mostly uniform city.

Hight - Density - Borderlines - Landmarks

Density

Even the physical density of built volume in the area is quite low, there are several areas of dense urbanity, which make the areal white inhomogeneous and generate, especially in the mind of high-rise, varying views and perspectives.

The felt density is in close connection to height and generates a chaotic picture of seeming independent structures put together in a seeming uncalculateable network, characteristic for most parts of Seoul.

Borderlines

Itself defined by very strong borders in the form of streets, the site's whole closer area is cut by capital borders, mostly in the form of streets, which split the homogenity and create pregnant borderlines defining the adjustment of the area. So even the site itself is strongly formed by these lines.

The only really dominant borders except the streets are the river's bank and the border of the military areal, supposed to be a park in 2050.

Hight

As a city with a great number of high-rise-volumes and spread mountains even in the city's area, views are likely to vary very strong. To the view is blocked by a giant wall of flat-structures, while to the northeast and northwest single block-bunches even give free to green mountains in a widely distant view.

In fact caused by the size of the volumes, these depend strongly on hight and position of viewer's perspective inside the areal.

Landmark

Because of its position in the center of Seoul, it's surrounded by dominant and marks as elementary points of orientation. Even they are not primary visible, they define the city's orientation the position of the site within. In consequence they seem to be essential ankers for the situation of the project inside the city.

Program

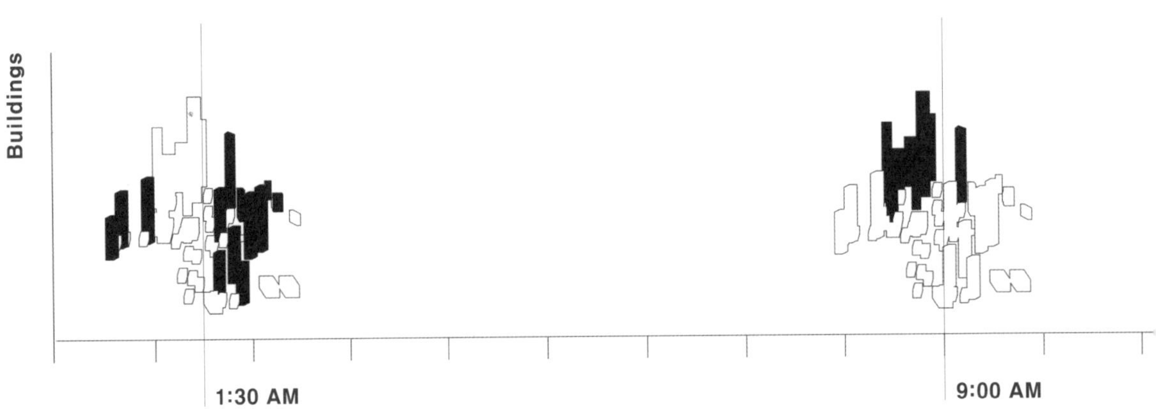

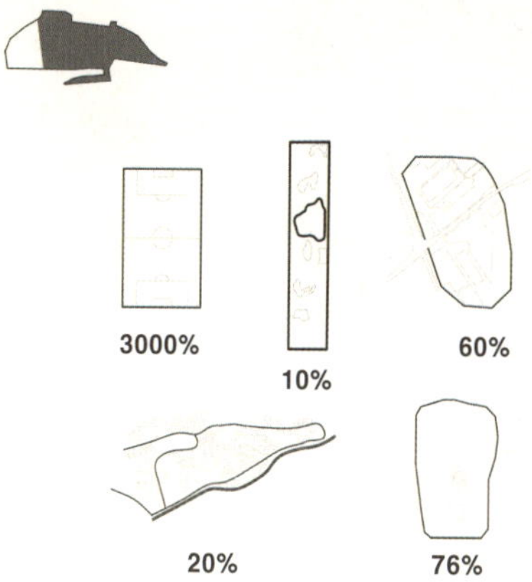

Studio Daniel Libeskind has been named winner of the master plan competition for the Yongsan International Business District of Seoul, South Korea. The new large-scale district features a number of residential, office and retail neighborhoods in an extensive urban park along the river.

Our group was commissioned to analyze the project, in order to scale the project that we were going to do during the semester.

The first thing that we wanted was to know which type of project we were facing. To do this we decided to compare the size of the project with buildings or sites known to everyone to understand each step of the project. The areas that we take as reference were: a football field, central park, le parc La Villette, Seoul National University, Gwanghwamun Palace. We compare the surface of the site and the project area.

Then we went over in detail with the program and explored the area in terms of site equivalence in football fields, movement and location Libeskind poses. In order to see how the project worked throughout the day we made a 3D diagram ticking the areas that are activated depending on the time.

We also analyze the transport network that connected the most important points to Yongsan and calculate the times that are estimated for each.

After knowing the scale of the project we realized that was very long and started studying which would be the impact of this project in the city of seoul. Some of the findings were as follows: The project is proposed for a population equivalent to that of Jeju Island, the tallest building in the plan will be the new landmark of the city as it is higher than the N tower.

With all this information we did the whole class an idea of the scale and operation of the project that arises for Yongsan.

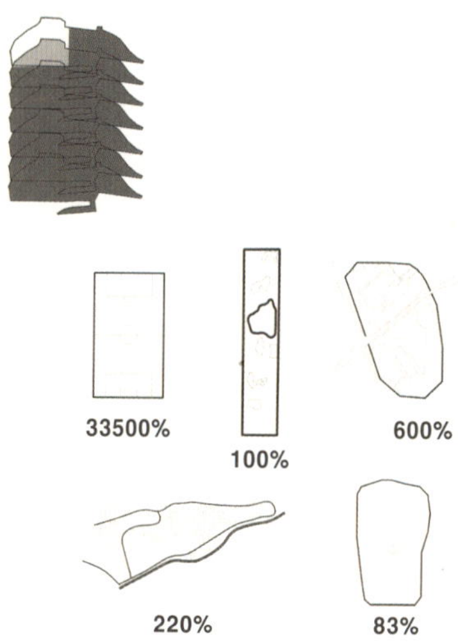

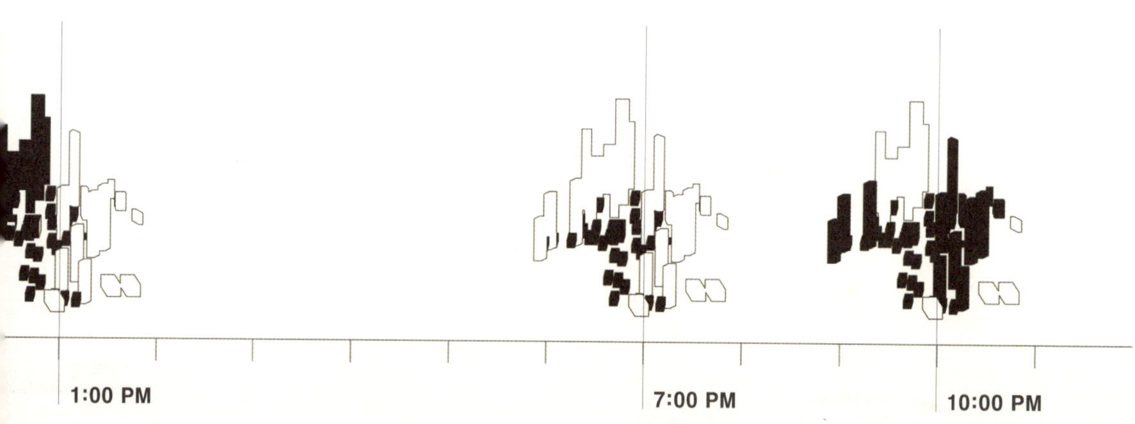

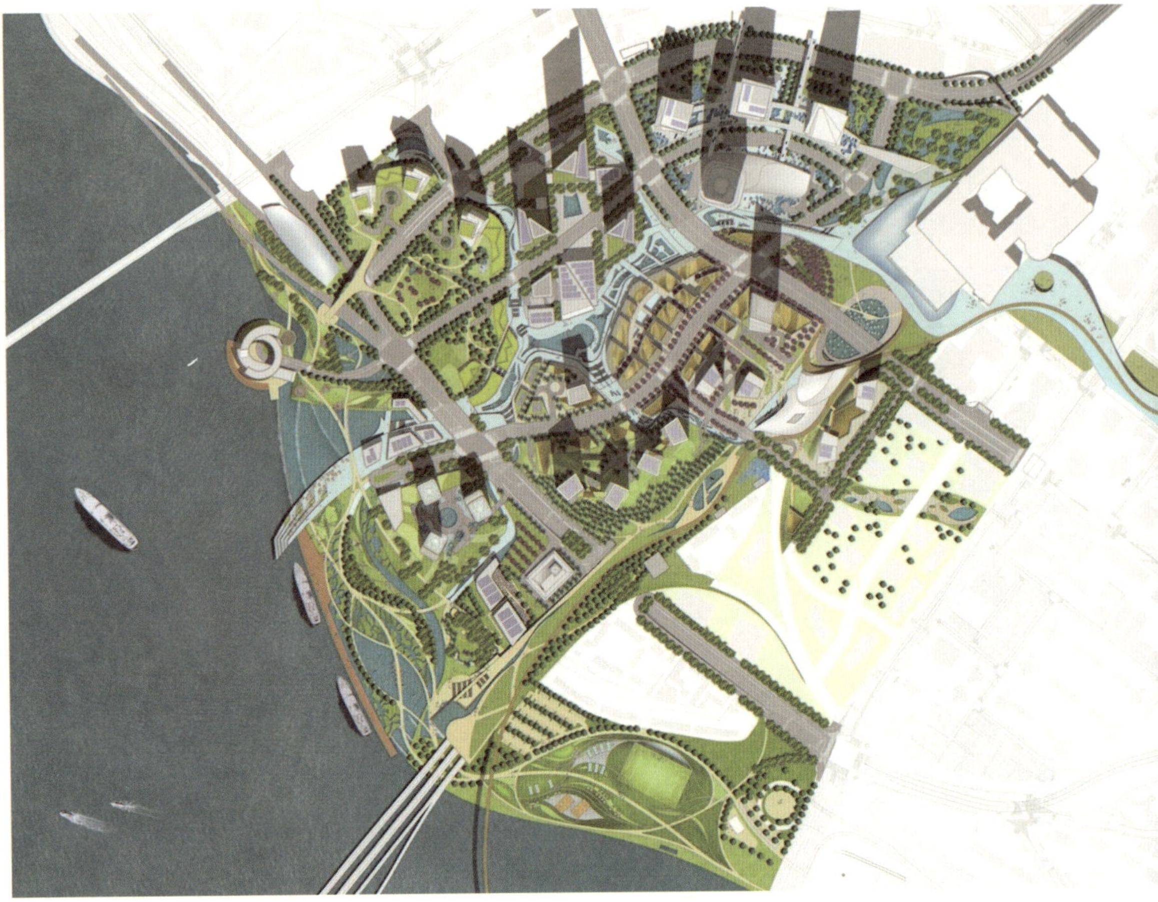

Future plan of Yongsan

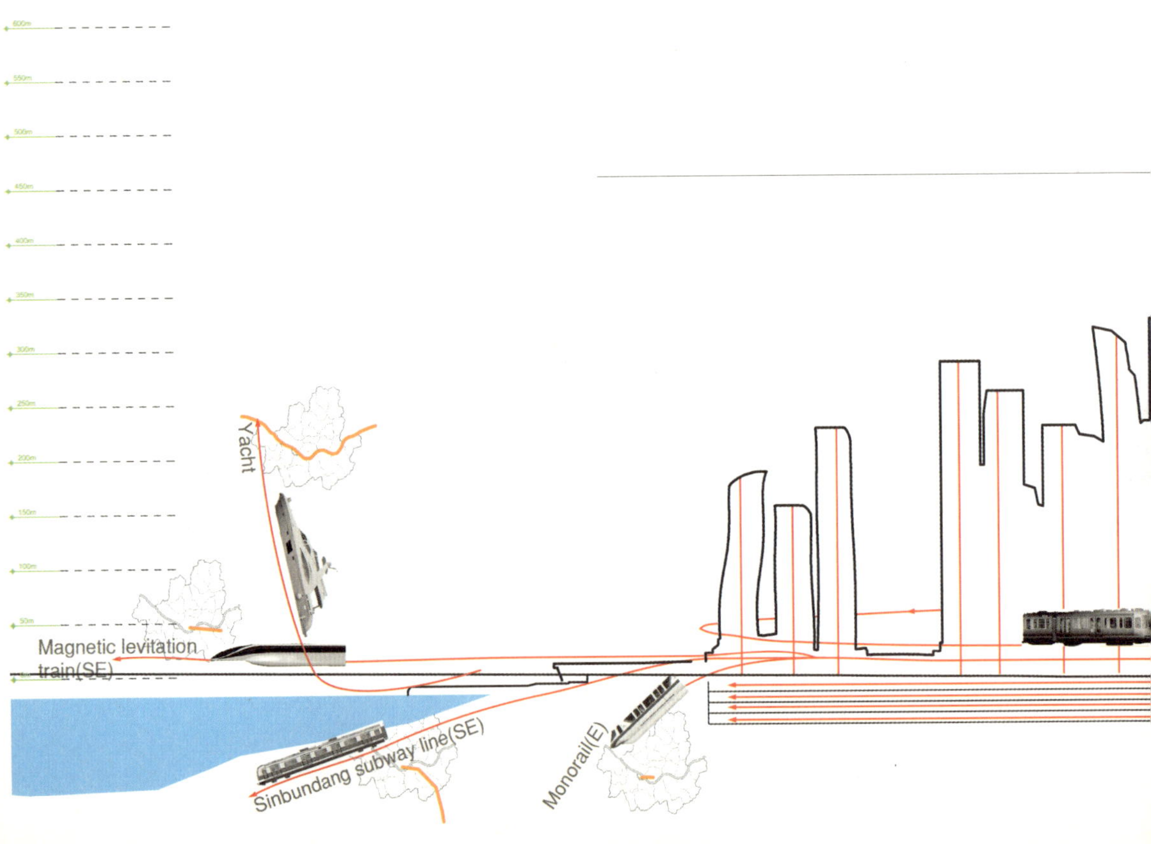

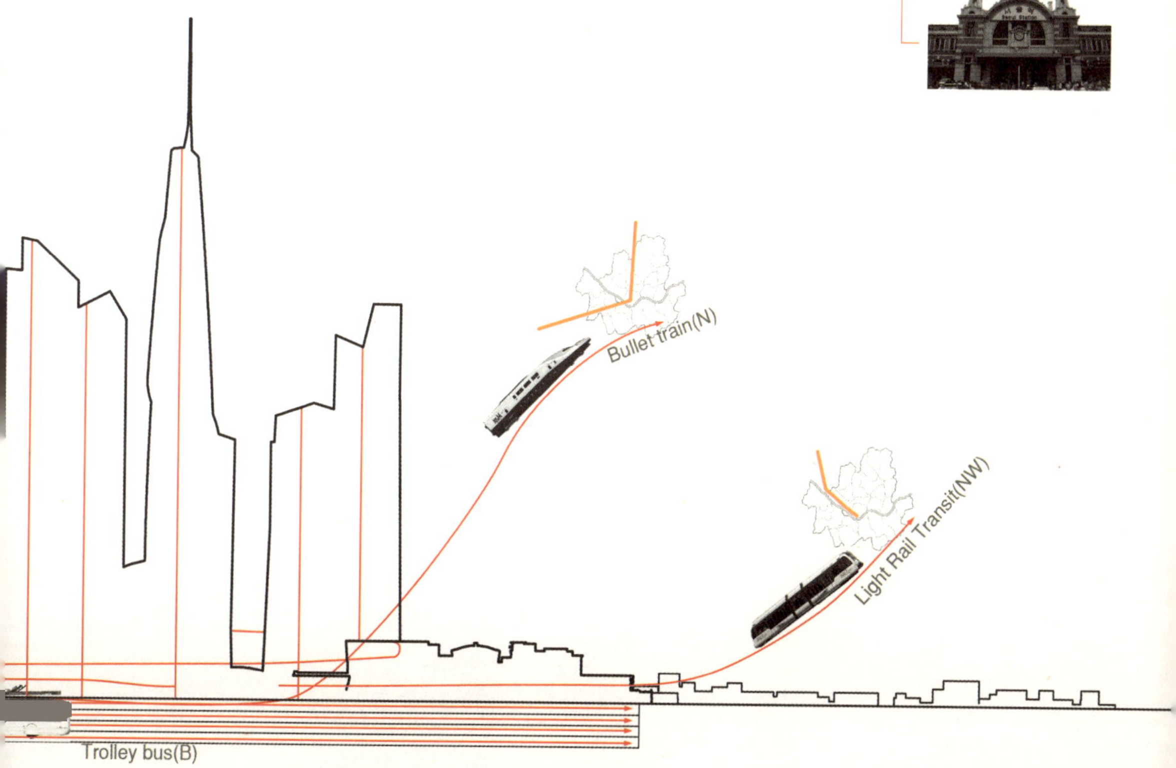
Bullet train(N)
Light Rail Transit(NW)
Trolley bus(B)

populations

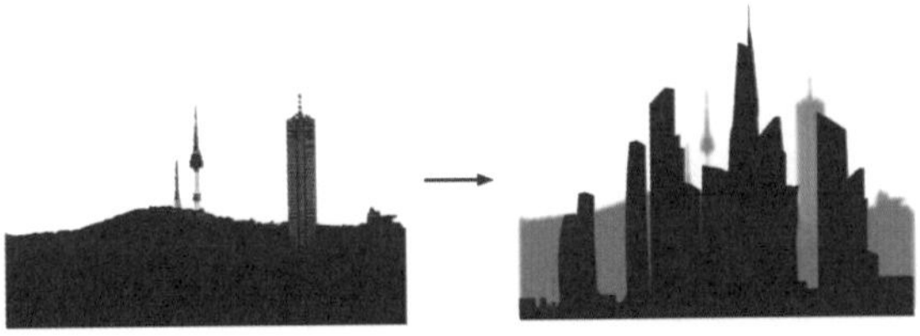

landmark

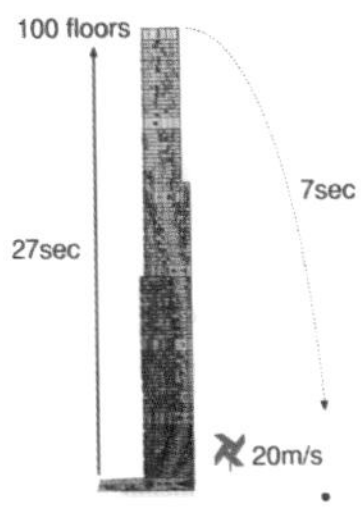

numbers

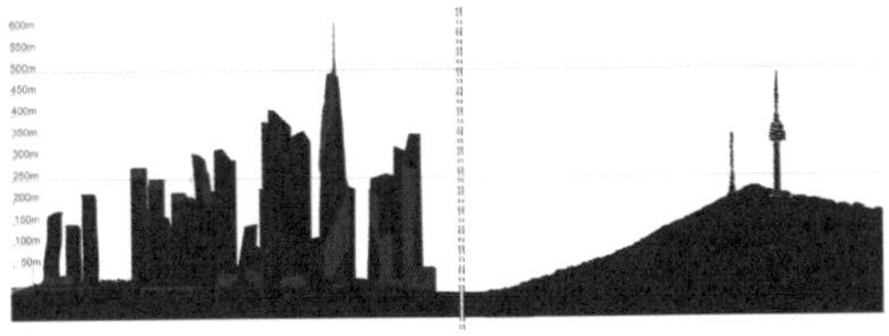

scale

Height of YIBD

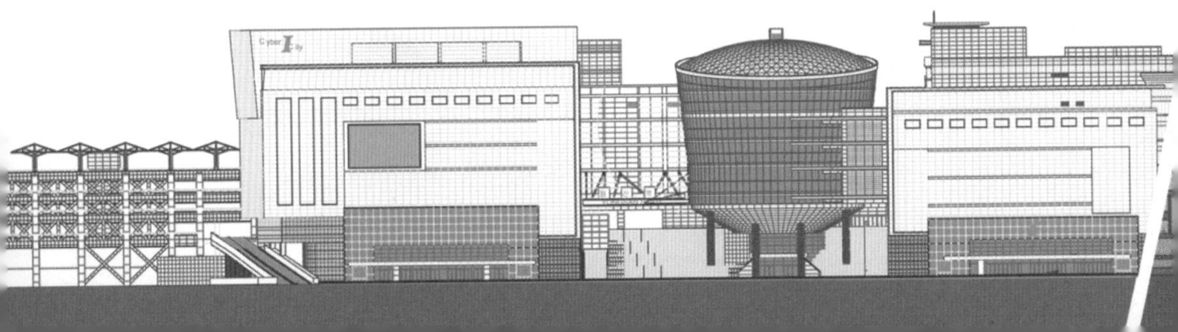

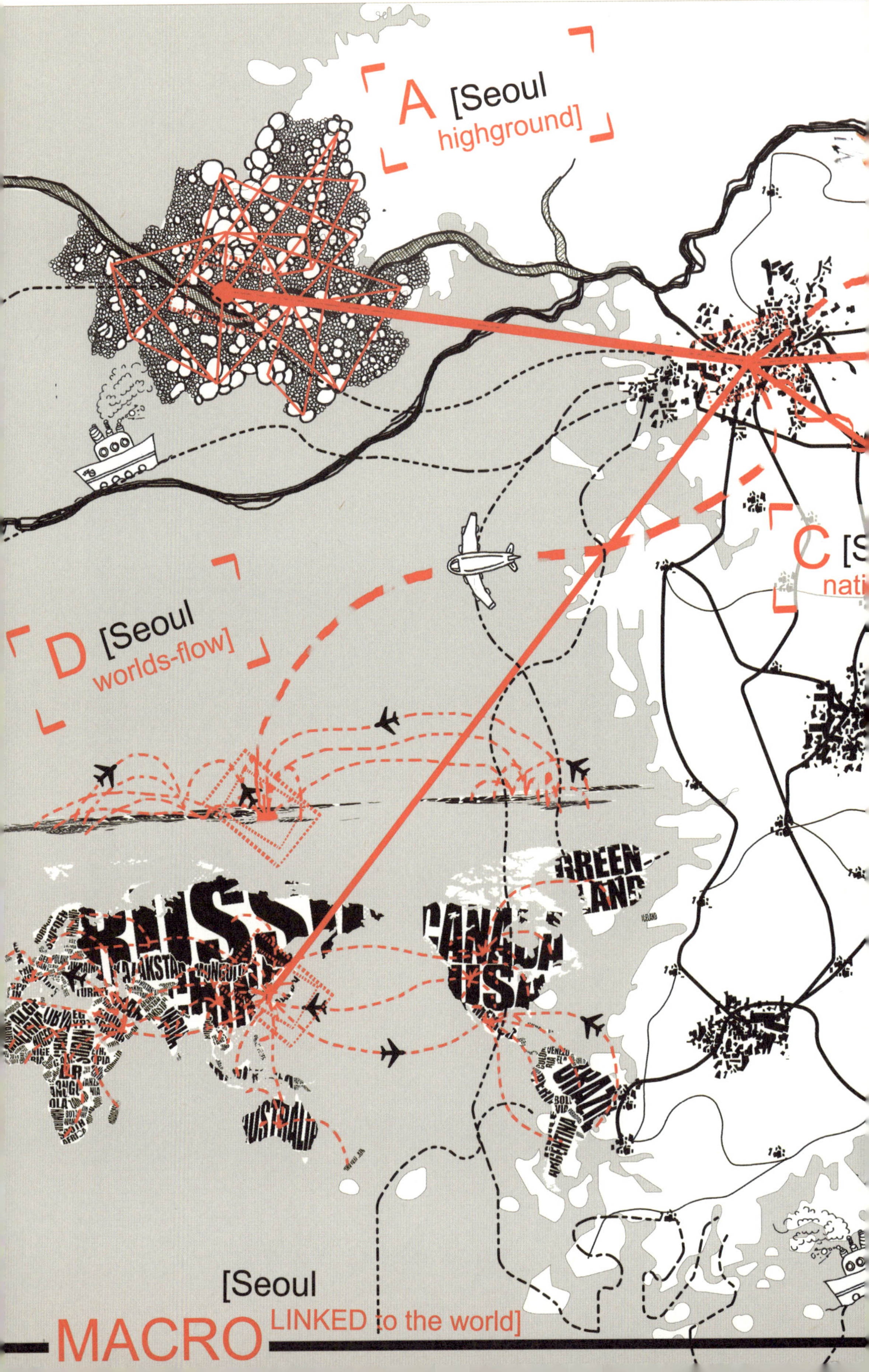
A [Seoul
highground]
C [S
nati
D [Seoul
worlds-flow]
GREEN
LAND
AUSTRALIA
[Seoul
MACRO
LINKED to the world]

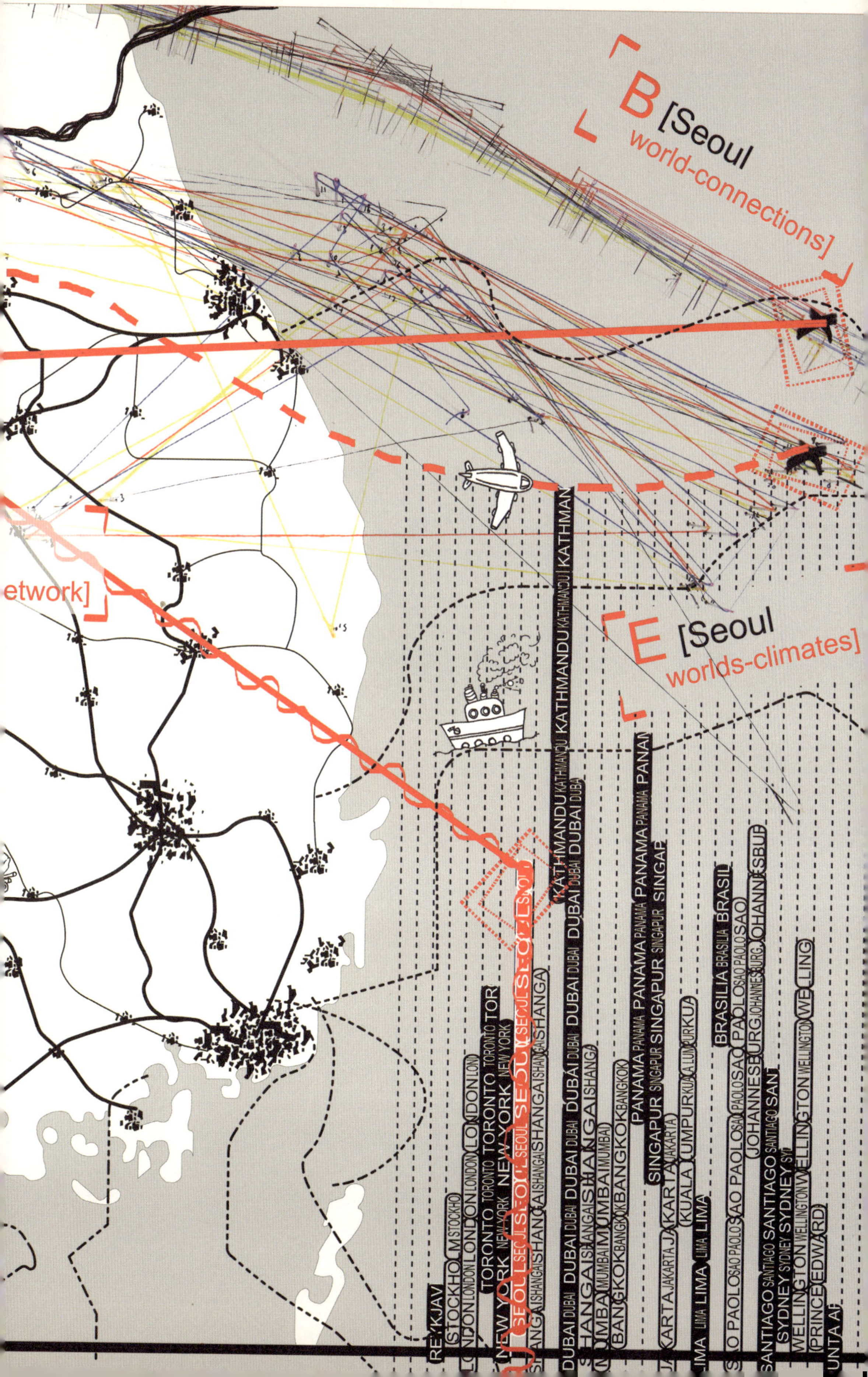
B [Seoul
world-connections]
etwork]
E [Seoul
worlds-climates]

PLANNING DREAMS OF SEOUL

“What Seoul Want to be?”

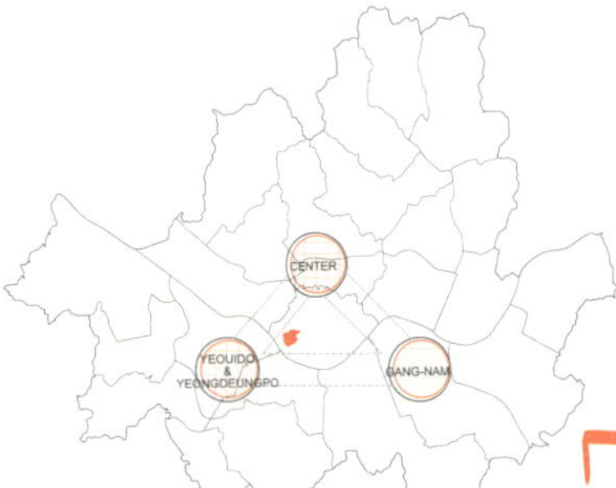

3 core

3cores 3subcores 13pots

main developing direction

A “Liveble-Global-Green Seoul”

Seoul City Plan(2030)

3 subcore

plan of region

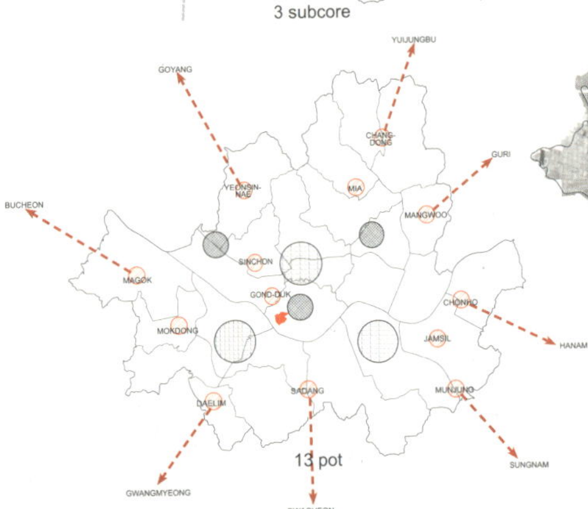

13 pot

C “Hard city to Soft city”

Design Seoul

scenery management area

water-front axis scenery area

design guide line :
water-front axis scenery area

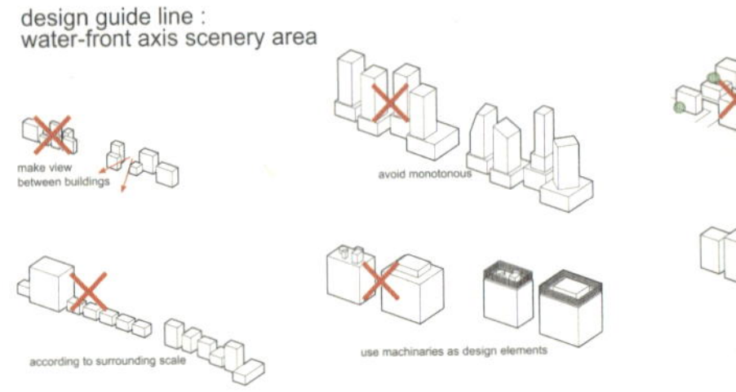

B "Create New Namsan Culture"

Namsan Renaissance

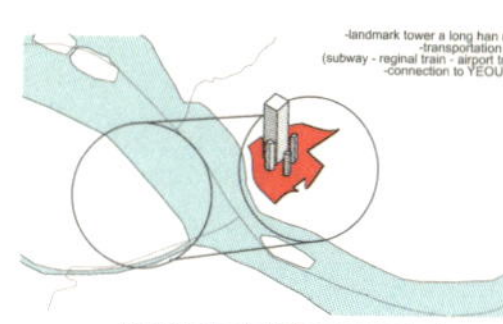

yongsan developement plan

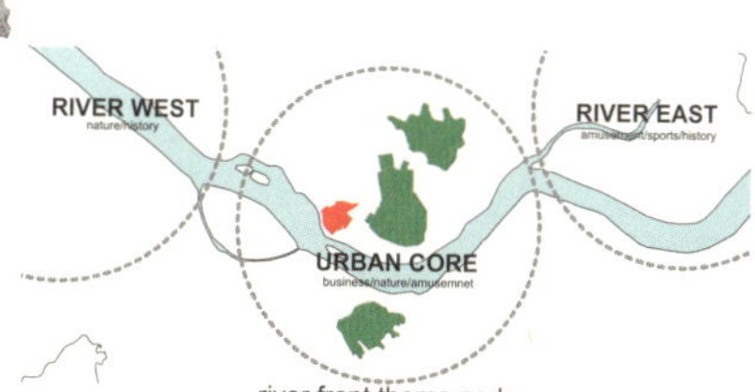

river front theme park

D "A World-class Attraction for Citizen & Tourists"

Hangang Renaissance

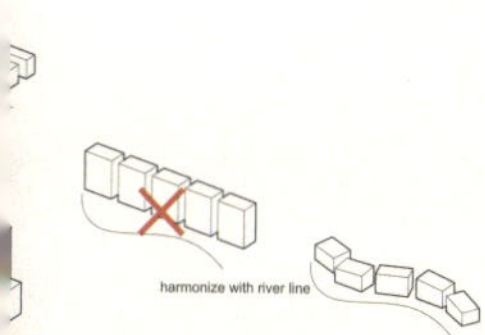

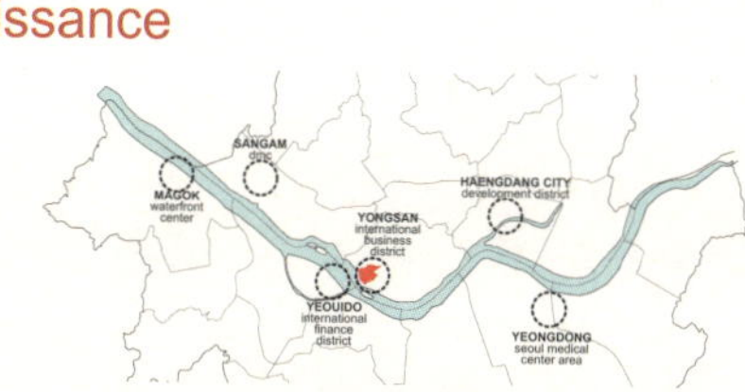

strategic centers

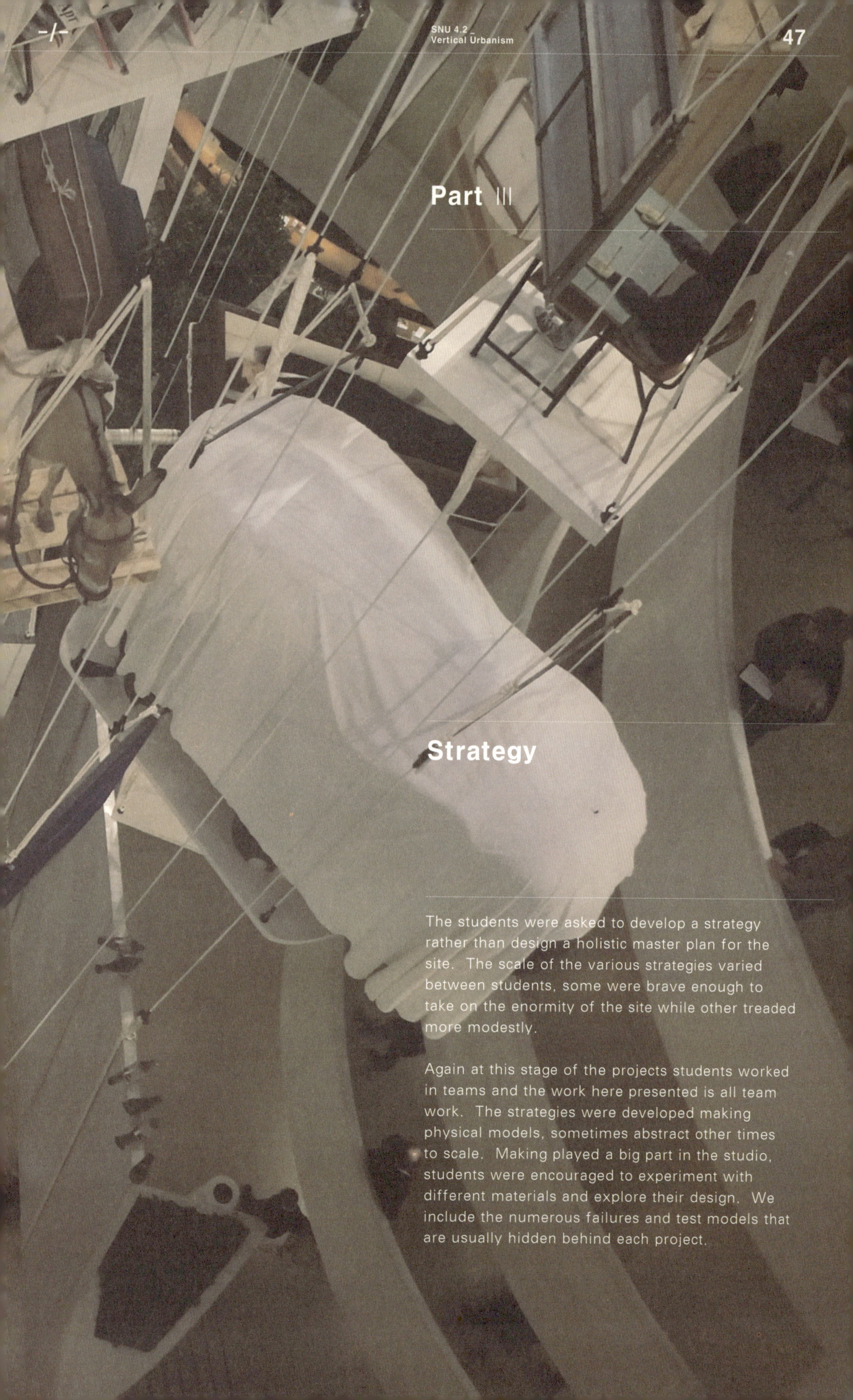

Part III

Strategy

The students were asked to develop a strategy rather than design a holistic master plan for the site. The scale of the various strategies varied between students, some were brave enough to take on the enormity of the site while other treaded more modestly.

Again at this stage of the projects students worked in teams and the work here presented is all team work. The strategies were developed making physical models, sometimes abstract other times to scale. Making played a big part in the studio, students were encouraged to experiment with different materials and explore their design. We include the numerous failures and test models that are usually hidden behind each project.

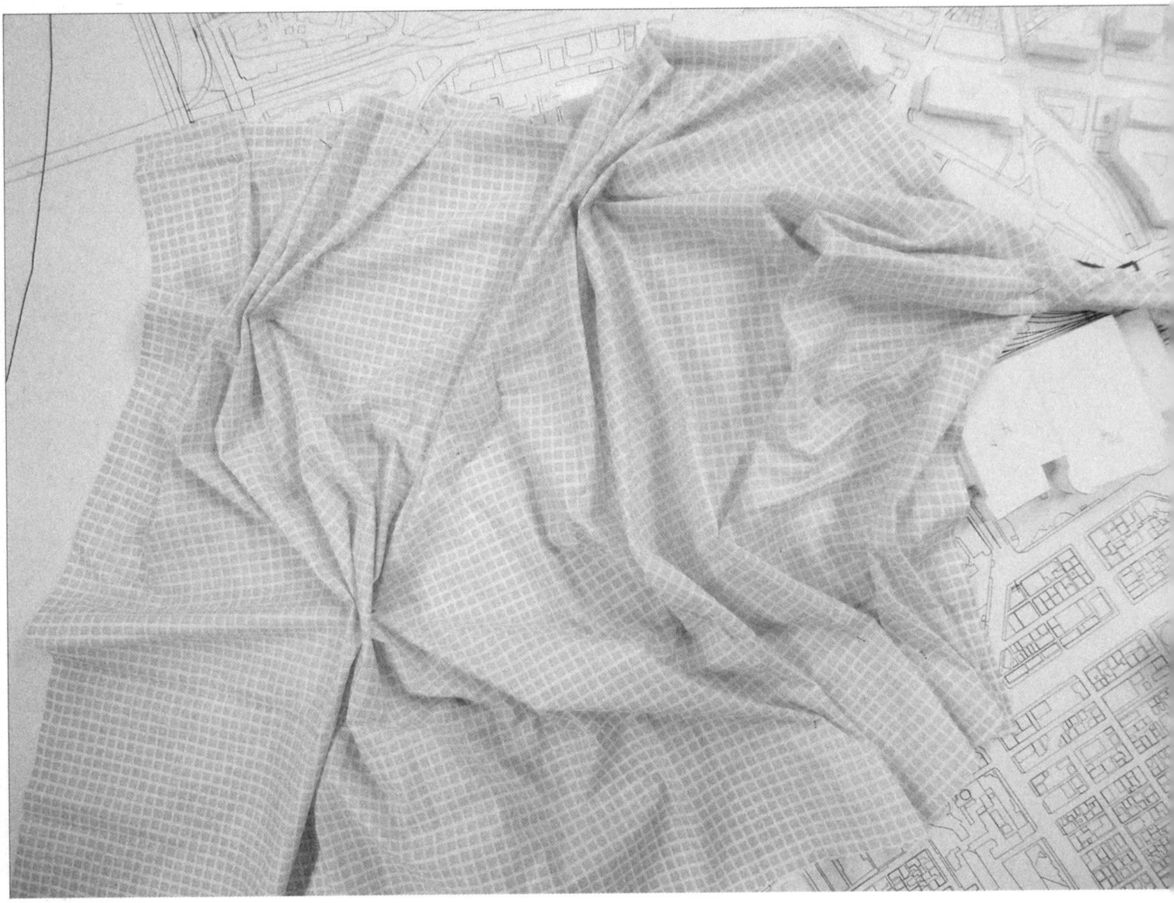

Team 1

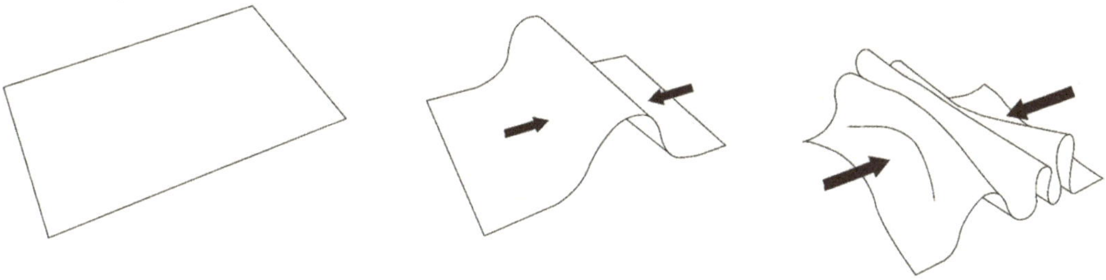

Wrinkles

Yongsan is the center of the transportation to the North Seoul and the gate to the old Seoul urban core, so it could have been far developed from Josun dynasty, Japanese colonial era to the present. Yongsan has a lot of historical traces. Urban grid system, modern time traffic facilities, old apartments, and new highirise residential buildings make the historical spatial layer, and the layers are entirely stacked here.

Our masterplan about Yongsan is not just filling the suggested programs fully in the site, or just placing the fancy buildings. Our strategy is making the WRINKLES of the fabric named 'Yongsan'.

Yongsan already has a lot of threads, flow, and density. We considered that high density points means where the wrinkle was made(on), and low density points means where the wrinkle was flattened(off). With wrinkles, we can make various vertical heights easily.

We could remind several words from on and off.

on :
urban, activity, event, meet, joint, accident, dynamic, density, congestion, traffic, population, systemic, grid, order, direction, hierarchy, artificial, controled, structure, infrastructure, controlled

off :
public, rural, uncontroled, green, farm, fresh, water, land, flat, low, slow, smooth, park, play, rest

With these concepts, we tried to find wrinkles already existed in Yongsan. Those wrinkles influenced our site and we could made new wrinkles in the site.

In our wrinkles, urban and rural stuff are not divided sharply. They are on the same clothing, so they are connected organically. Our wrinkle city is about not only physical density, but also invisible urban conditions. It is vertical, and also organic.

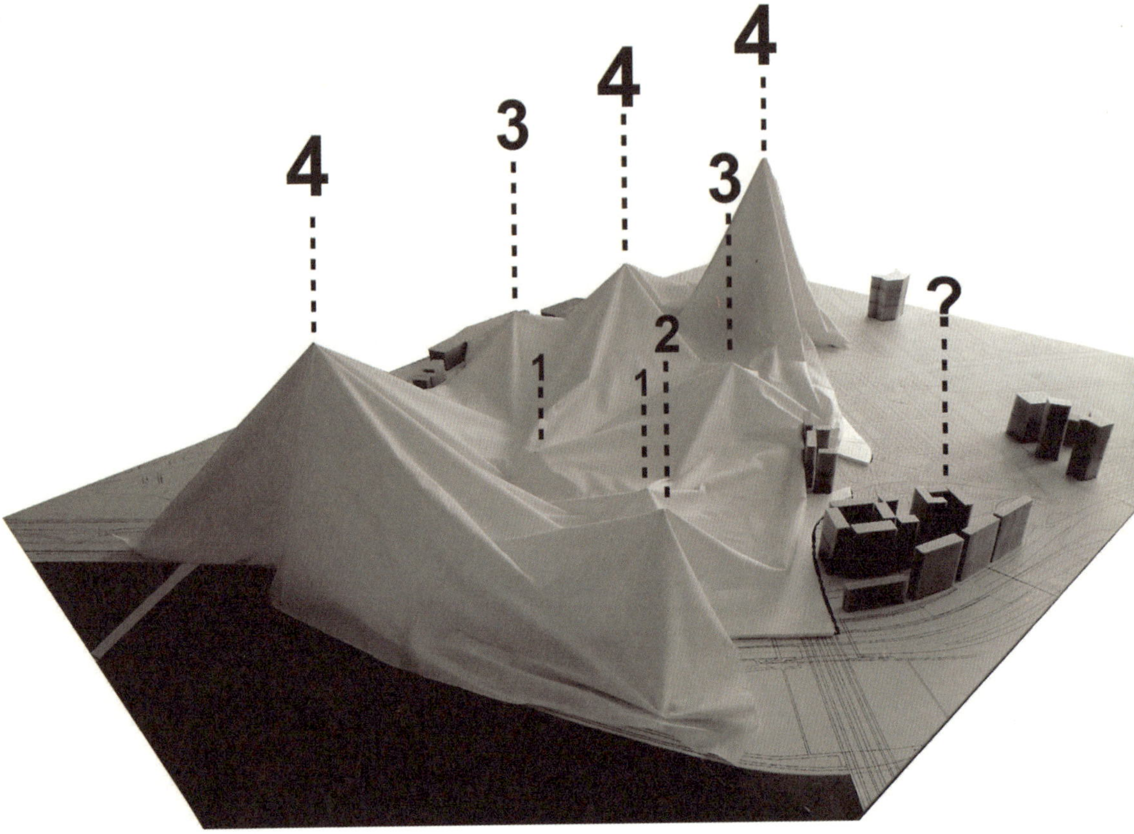
4
3
4
4
3
2
1
1
3
?

Team 2

We worked in group the master-plan research. The first approach was a proposal that took into account the context. It was a social proposal, whose base was the relationship between life and work. The problem with it is that it couldn´t develop in vertically. Therefore we decided to discard it.

After that ,we were decided to do a wide variety of proposals about how we could made the master-plan. For example: Put all the uses in a group in a building or organizing the uses in parallel lines occupying the whole site... We are left with a proposal to build in height , the main idea of this was trying to leave the ground free.

When we had the general idea, we begin to see how it could be real architecture. We studied different types of structures to make the idea to be transformed into architecture: a mega structure whose floor support was minimal, and the other option was to use the minimal structure but with more points of contact on the ground.

We chose the second type of struture because it was most real and functional than the first one. Once we had got the structure, we began to see how we could do the building. we began to see how we could do the building. We did not want a closed building, we wanted to a porous building, where light can pass through the building.This would solve the shadow problems on the skyscrapers.

With these main ideas we decided to make several approaches, the first was a poetic proposal, where we try to put the whole program together. To make this the program was stacking in a chaotic and random way. It was impossible to build. The model was an abstraction of what we wantedZ to do.

Then our next proposal was put in order ideas. We decided to climb the level of the street so the ground will be free.We wanted the main flow was at the top of the building thus changing the logical order of the program and activating movent in the building. So we decided to use the top for public and business space, and the bottom for residential.

The main ideas were to up the street to leave the ground free, and thus recover green spaces and gardens that are in the suburbs. Because in these cities, as large and constantly changing as Seoul, this areas have disappeared. Another idea was make a unit of city, where you can live and work, without feeling overwhelmed to live in the city center.

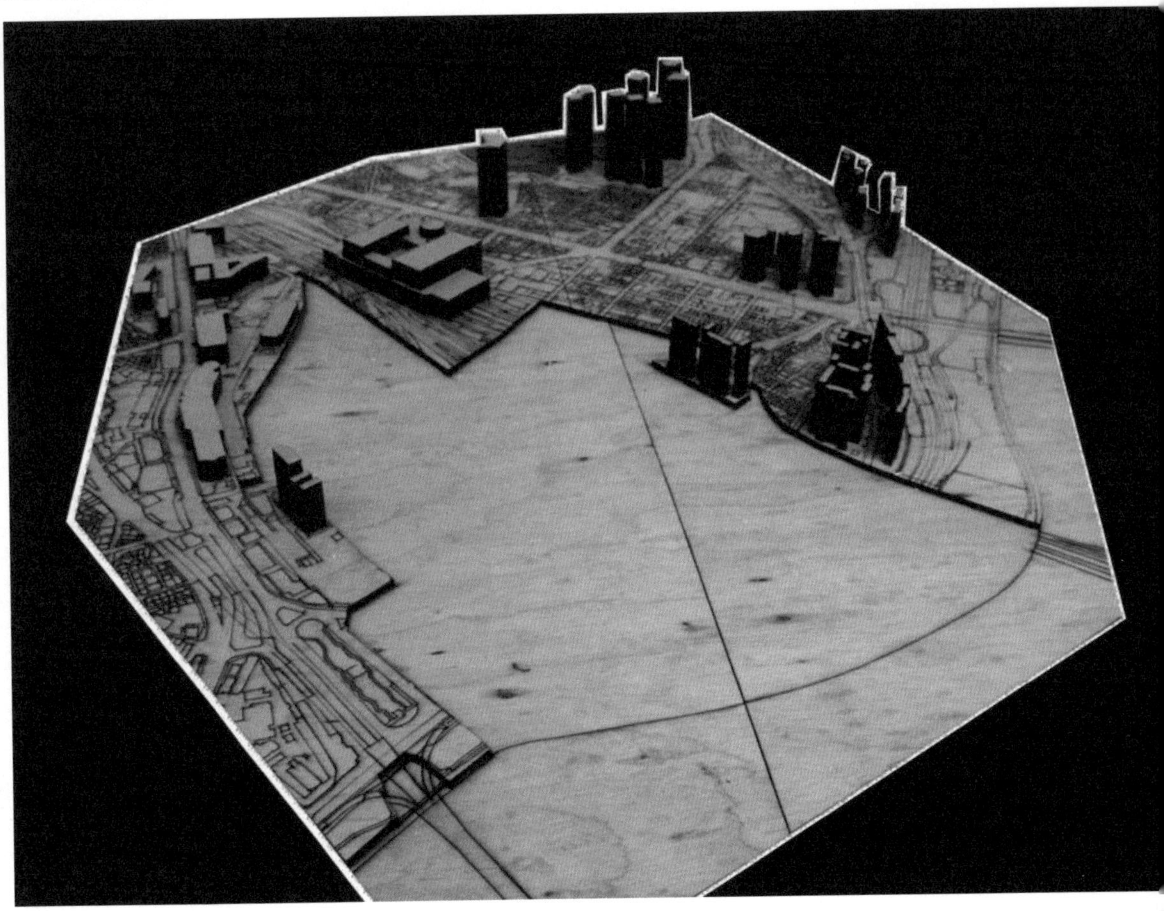

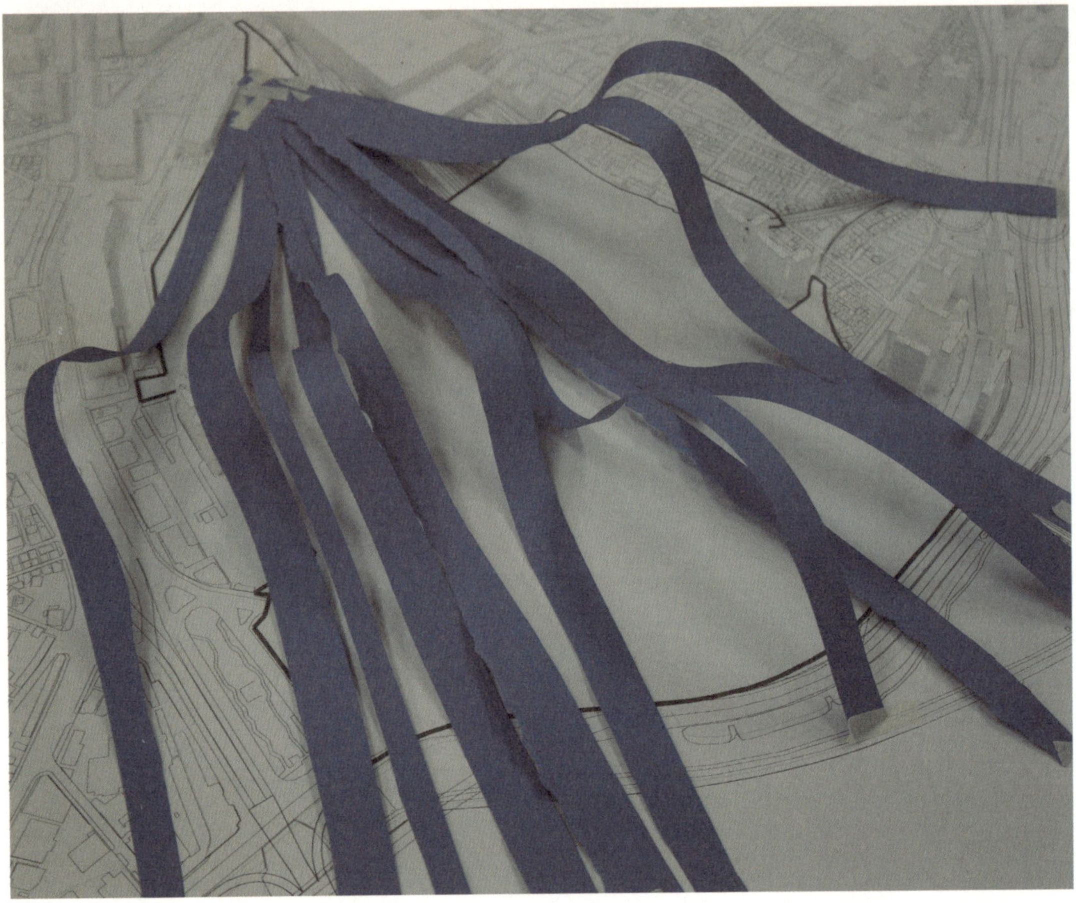

Team 3

Site is loacted on Yongsan train station. So it seems like a section of napa cabbage and the section can embrace may layers like Kimchi. After some studies, horizontal layers was developed to vertical wavy Kimchi. It can be a city or a building, depending on scale. And waves can cross each other.

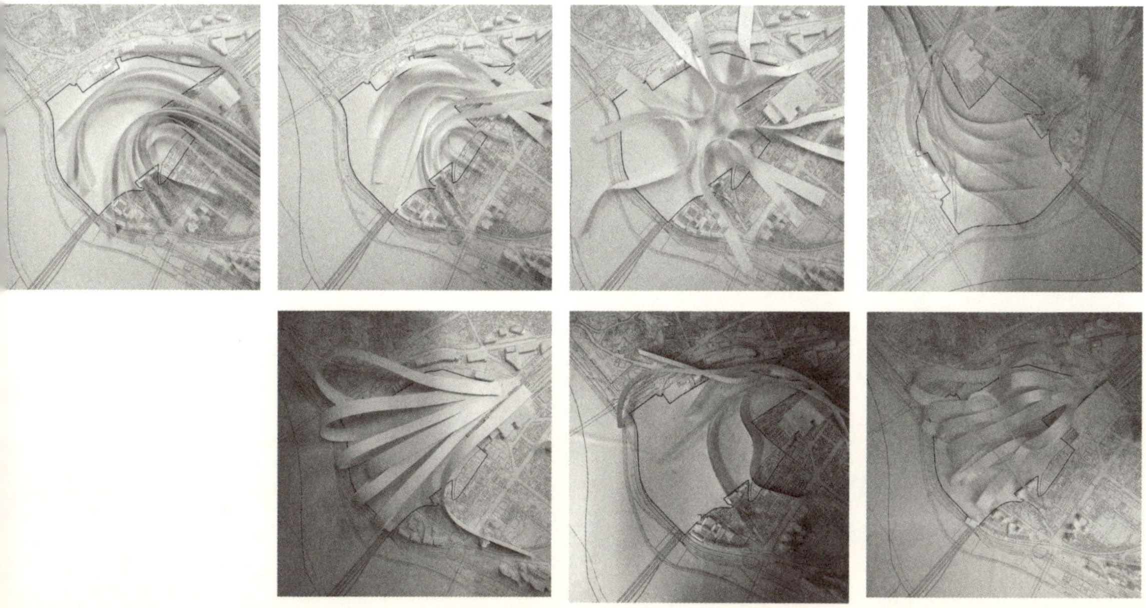

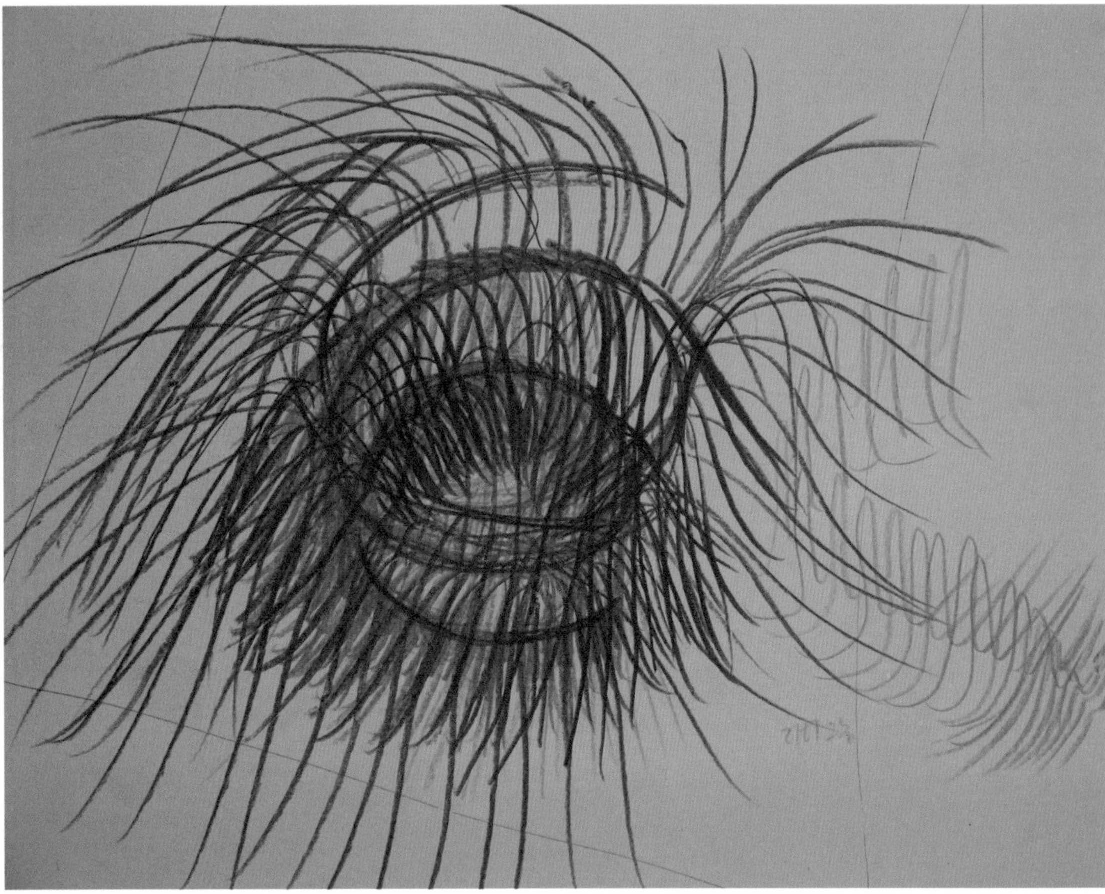

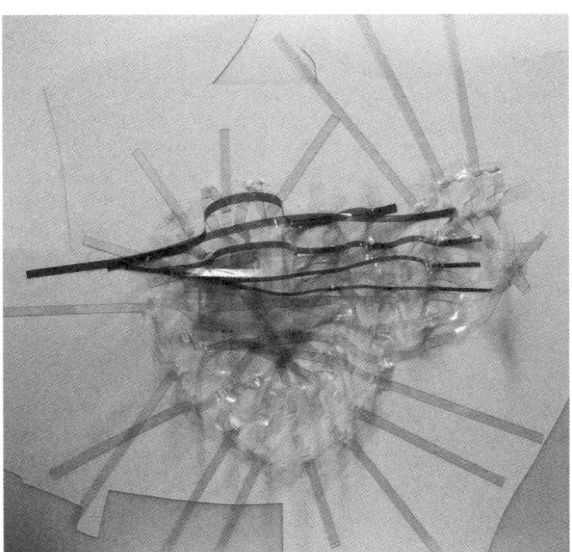

Team 4

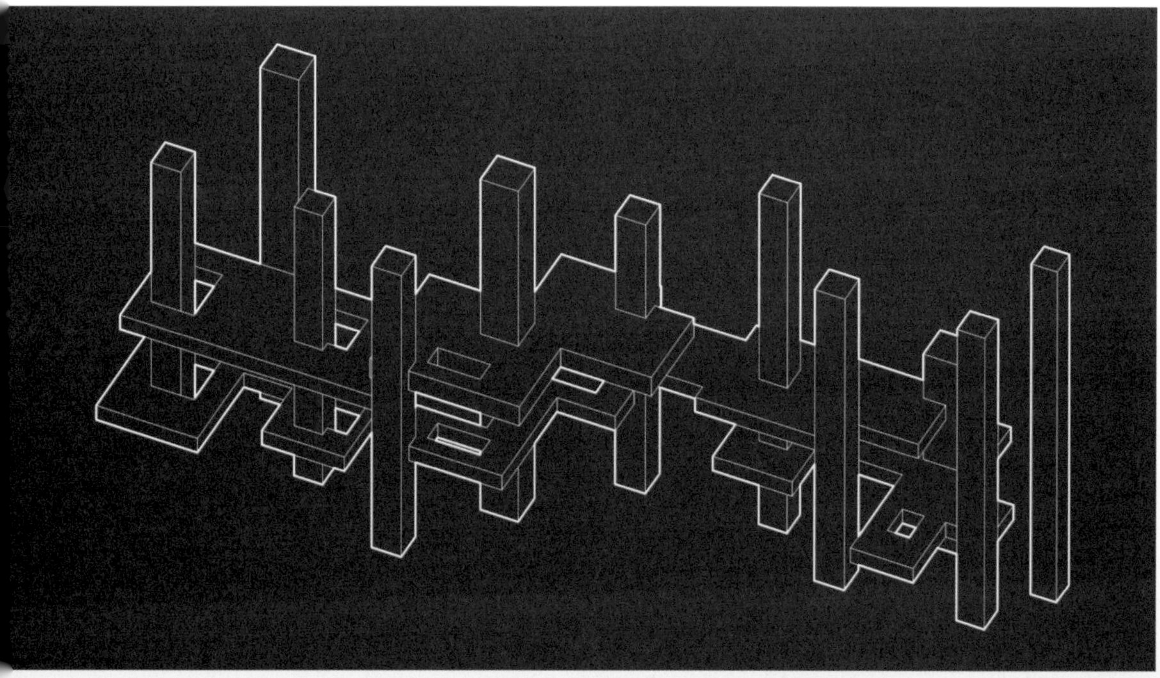

Physical connection between buildings

Because of accessibility, present connections of buildings are just lower parts of buildings. And these parts are used as public spaces and upper parts are used as private spaces. In master plan this public parts can be lifted up to make new space in the towers cluster. By doing like this, it's possible to have more usable space on ground level and additional access level to the buildings.

Furthermore, if more than one plate are added by stacking, towers and plates are become cores and slabs of one building. And when put different plates, various spaces are created between towers and plates. And with these plates, tower could be a connection joint. Thus, this formation of towers and plates could be a new type of building composition.

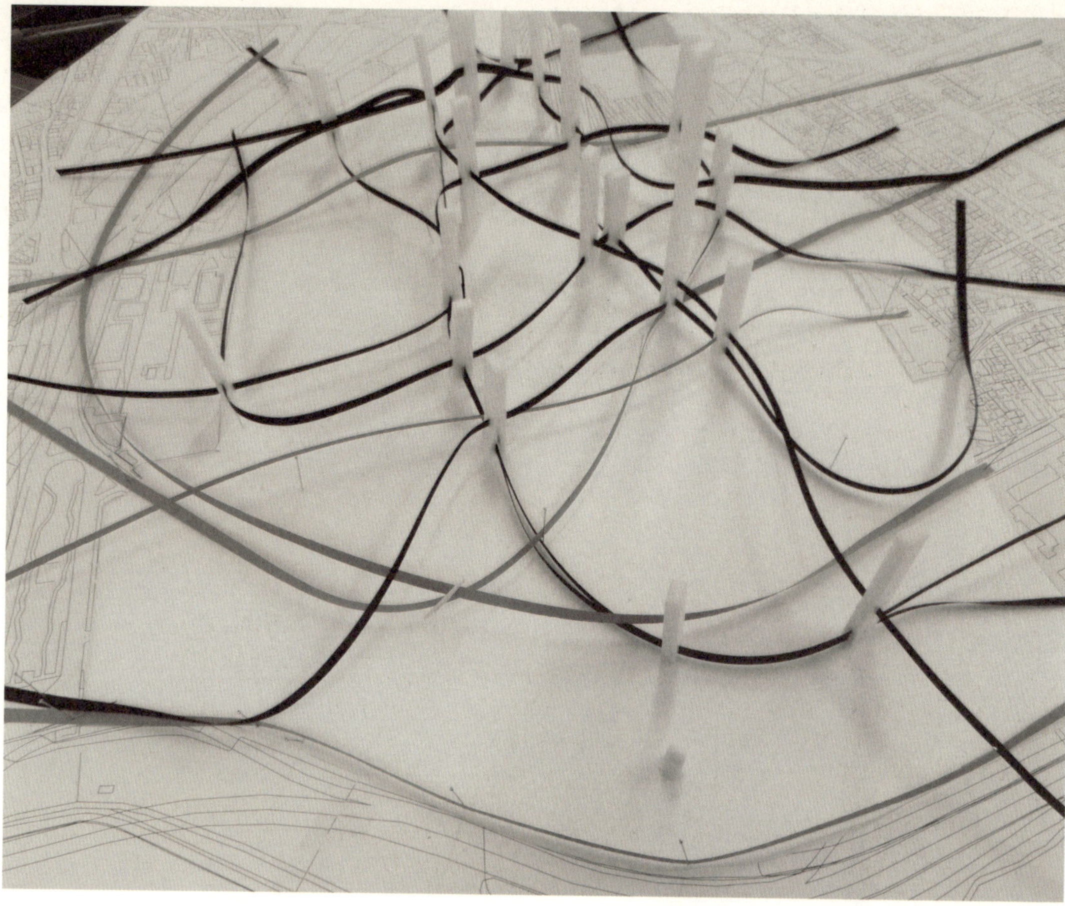

Team 5

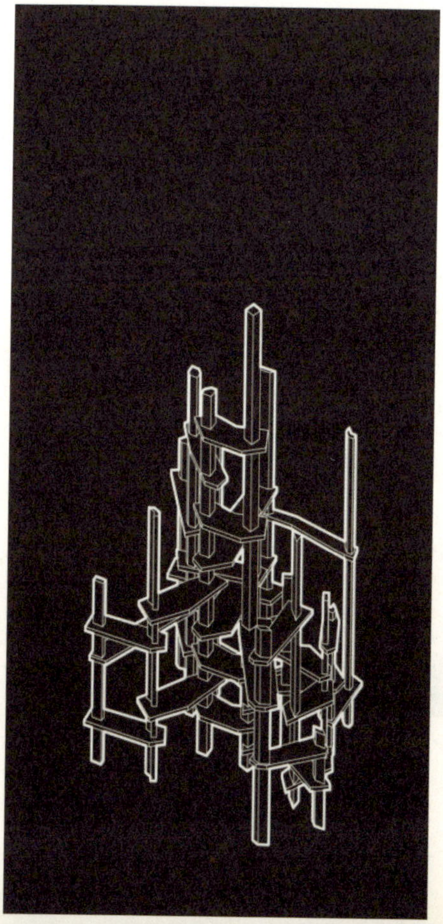

Rule of game

Every city has their own organizing method, like a game rule. And it's make sense that the way designing a master plan could be making a rule of game. In 'YIBD', we tried to apply game rule.
Basically, a city have road first, and with this road network, divided surfaces are used as building areas, like 'Checker'. 'Checker' is a board game, in game, stones are placed on a square which is divided by grid. So, we tried to have a reference on 'Go'. Similar game of 'Checker', but when people play 'Go', they put stones on the junction of the grid.
In site, we can make any kind of grid, and on grid, we can put buildings on the junction of the roads. Usually, Buildings are working as a solid of city, but in 'Go' system, each buildings have included roads in themselves and working as a road and a junction too

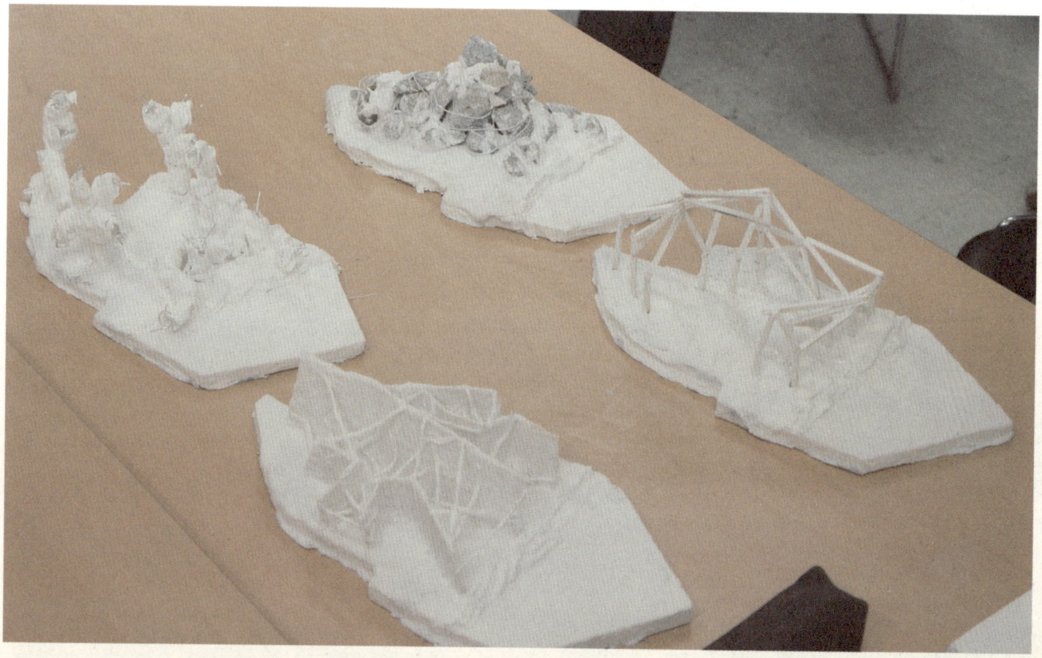

SEOUL NATIONAL
UNIVERSITY REVIEW
Fall FINAL Review
Lucia, Espinoza de los Monteros
Duk Hwan, Kim
David, Kriechmair
Seok Min, Ahn
Hansol, Choi
Myoungki, Lee
Hyejin, Song
Do Hyun, Baek
Han Chan, Ryu
Marina, Kinderlan
JURORS
Young Kyoon Joung
Peter Winston Ferretto
Ali Rahim
Evan Douglis
Irina Verona
Helene Furjan
Peter Trummer
Upper Gallery
Wednesday,12/14
10:00am

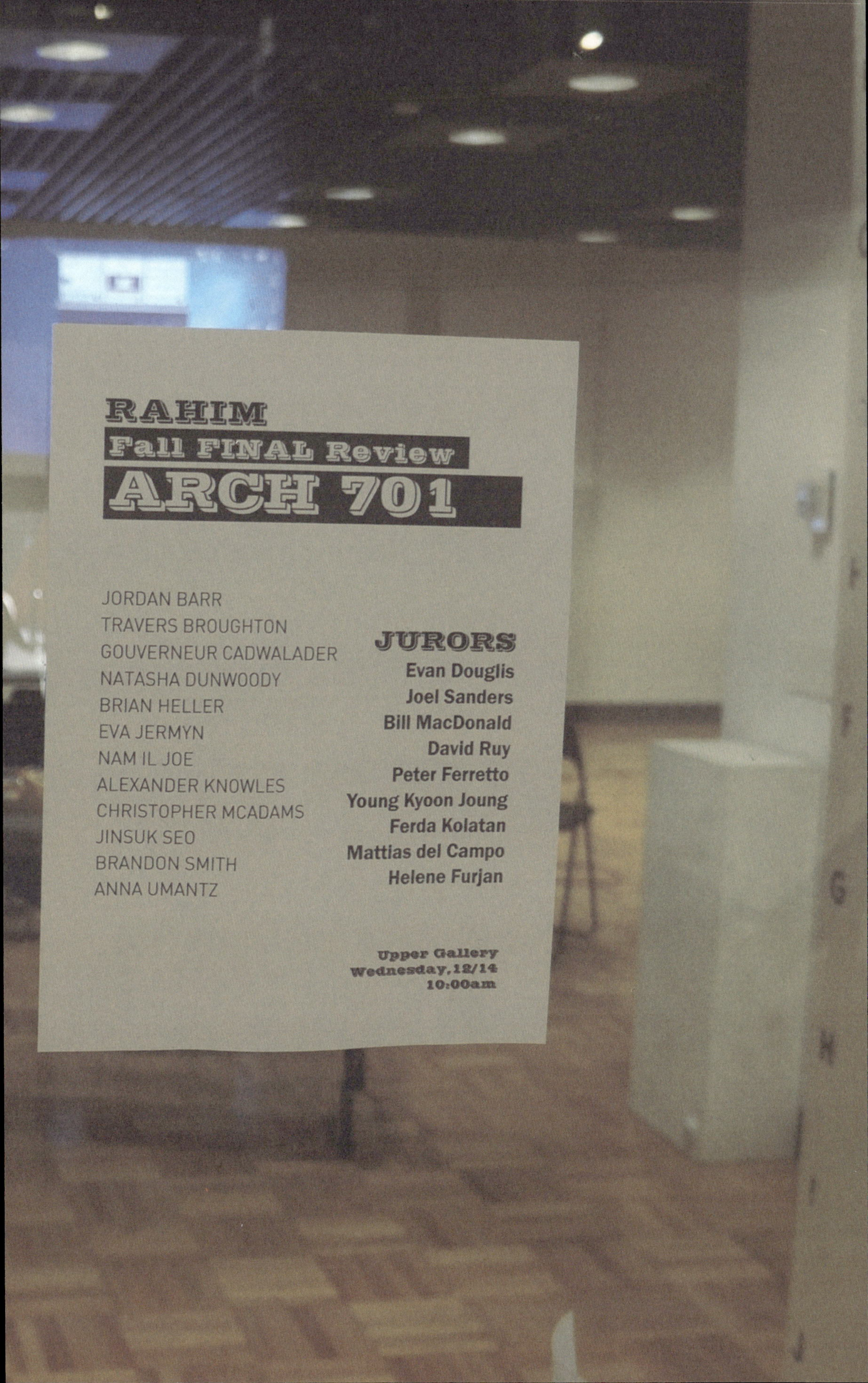
RAHIM
Fall FINAL Review
ARCH 701
JORDAN BARR
TRAVERS BROUGHTON
GOUVERNEUR CADWALADER
NATASHA DUNWOODY
BRIAN HELLER
EVA JERMYN
NAM IL JOE
ALEXANDER KNOWLES
CHRISTOPHER MCADAMS
JINSUK SEO
BRANDON SMITH
ANNA UMANTZ
JURORS
Evan Douglis
Joel Sanders
Bill MacDonald
David Ruy
Peter Ferretto
Young Kyoon Joung
Ferda Kolatan
Mattias del Campo
Helene Furjan
Upper Gallery
Wednesday,12/14
10:00am

Part IV

Projects

The final part of the design studio was for the students to separate from their teams and generate individual projects in the format a high-rise building. The 10 projects presented at first sight all appear very different, and this is a point that was strongly encouraged in the studio however on analysis the projects one can start observing common architectural thinking patterns.

Structure, public space, circulation became the lexicon with which the majority of the students developed their projects. The results and the atmosphere created by the stundets, both in Seoul and Philadelphia was remarkable. The studio will continue to pursue the theme of vertical urbanism in the future and hopefully this book is the first of many on the subject.

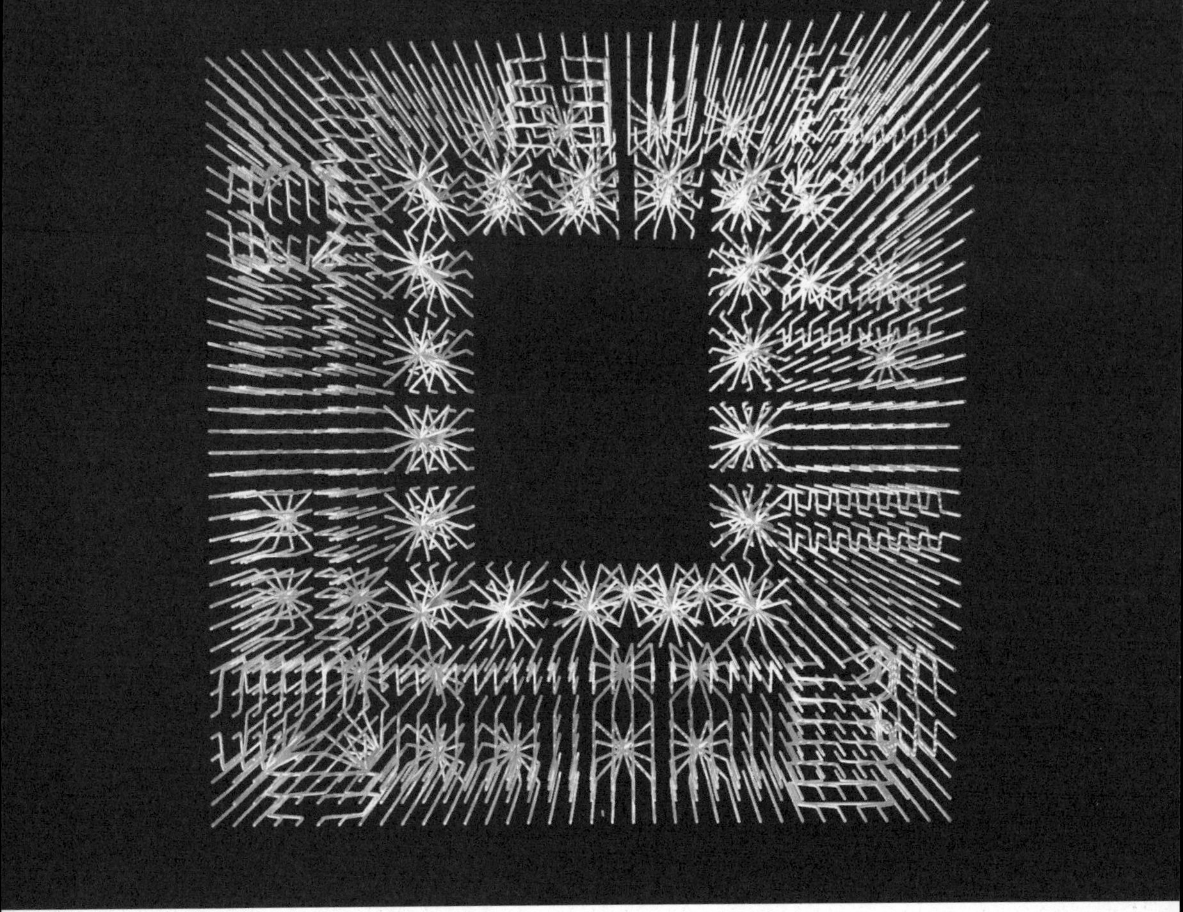

(Un)_Predictable Systems

Kim Duk Hwan
KOREA

This project is called 'un-predictable systems'. It is making unpredictable systems using predictable systems.

First phase of the project was for team work. Our project started by looking at the typical high rise towers and we made an opposite approach so to make the new form of sky scrapper, what we called 'anti rise'.

We looked at how sky scrappers actually work. And what we wanted to propose is shifting the public domain into the higher position of the sky scrapper (compare to the typical typology of towers' public domain at the bottom) therefore create a new public realm. And we focused on many ideas that breaking the general notion of sky scrappers. So we planned to make an exploded tower which had been a solid and monolithic so to improve the problem of shade, light and also repetitive plans.

And we completed the concept by making the adjustable, scattered structures on the regular grid, which rules the position of the columns.

And that rigid, regular grid was starting point of my own project in second phase. I focused on these grids which is quite determined but if i break it a little bit, I can get a lot of freedom on the contrary.

So I imagined a forest of columns in a very regular grid, then I tried to mutate it
that could become a new form of structure. Changing the rules by a little actions, we can extract many different types of spaces.

So we have a forest of columns in regular grid as a base structure. and then split, and grab or just release. Consider the structure as a framework. The spaces can be made up inside the structure depending on the use of it.

This simple logic can generate various types of spaces. You can see that how the column forests inhabited and you can also find some strange, unpredictable spaces like a valley, like a river crossing through the structure, intersecting public spaces, bunch of columns to make cores, and most importantly, columns adapting on the conditions of the surroundings.

By using those tools, we can generate not only a sky scrappers, but also a new form of adapting city which is unpredictable consequence, by predictable actions.

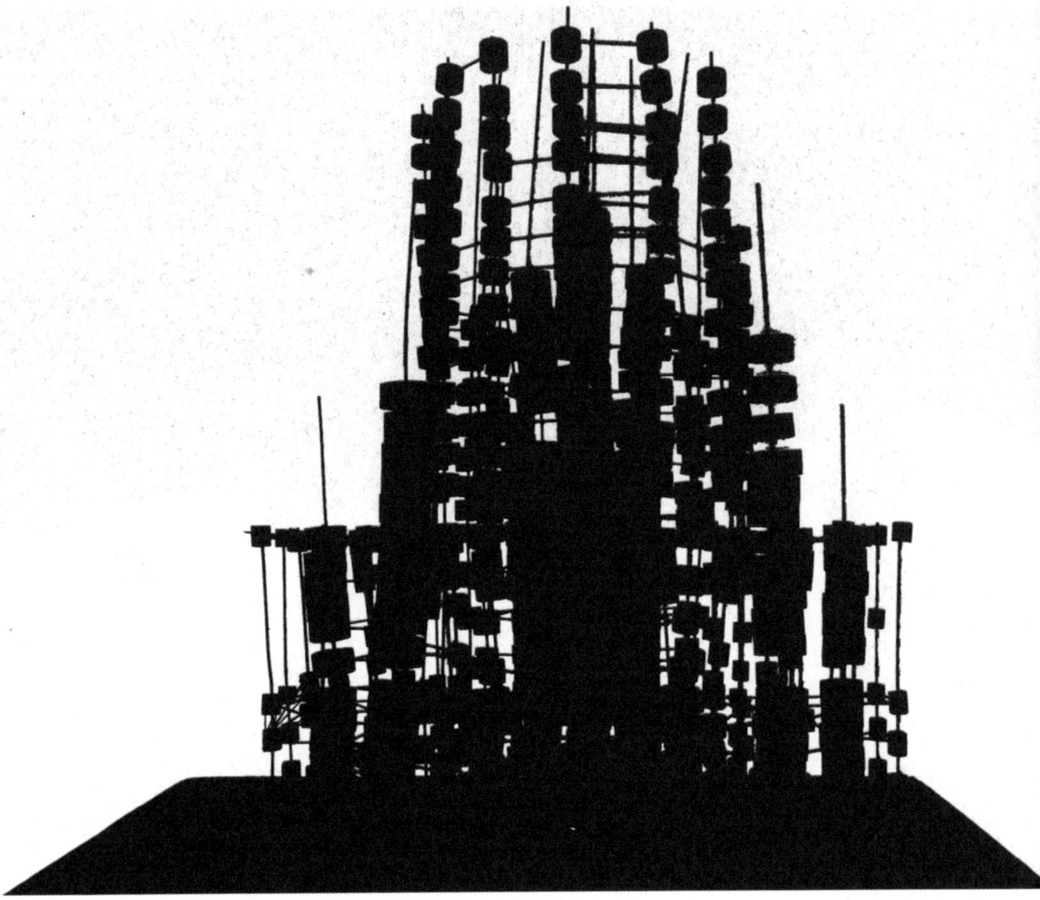

Relative complexities

Ahn Seok Min
KOREA

People is one of the most important thing about city. I studied activities of people. And classified them into moving, living, shopping, playing, studying and so on. From them, I selected some important characters of Yongsan like traffic, housing, nature, facilities. And I divided them into small subjects like road, train, apartment, high-apartment, green, water, military base, embassy, entertainment, school and so on. And I represented them on the map saperately and together. They can explain the situation of Yongsan and the spatial arrangement of Yongsan in some degree.

Site is loacted on Yongsan train station. So it seems like a section of napa cabbage and the section can embrace may layers like Kimchi. After some studies, horizontal layers was developed to vertical wavy Kimchi. It can be a city or a building, depending on scale. And waves can cross each other.

Beginning was subject of public & private in highrising buildings. I Thought that it depends on relative relation of size, distance, diversity, density, shape and so on. So I arrayed dots in many ways and got some plans. And depending on size, arrangement of dots, I give function to each dots. And I arrayed some plans vertically and connected them. After some studies, I made base plan which has a basic composiotion and a posibility to be changed. And I piled 100 stories with 3 different cylinder modules to 2 meters high (diameter of cylinder: 20mm,30mm,50mm). And to represent a possibility of waves crossing, I use two white, red cylinders.

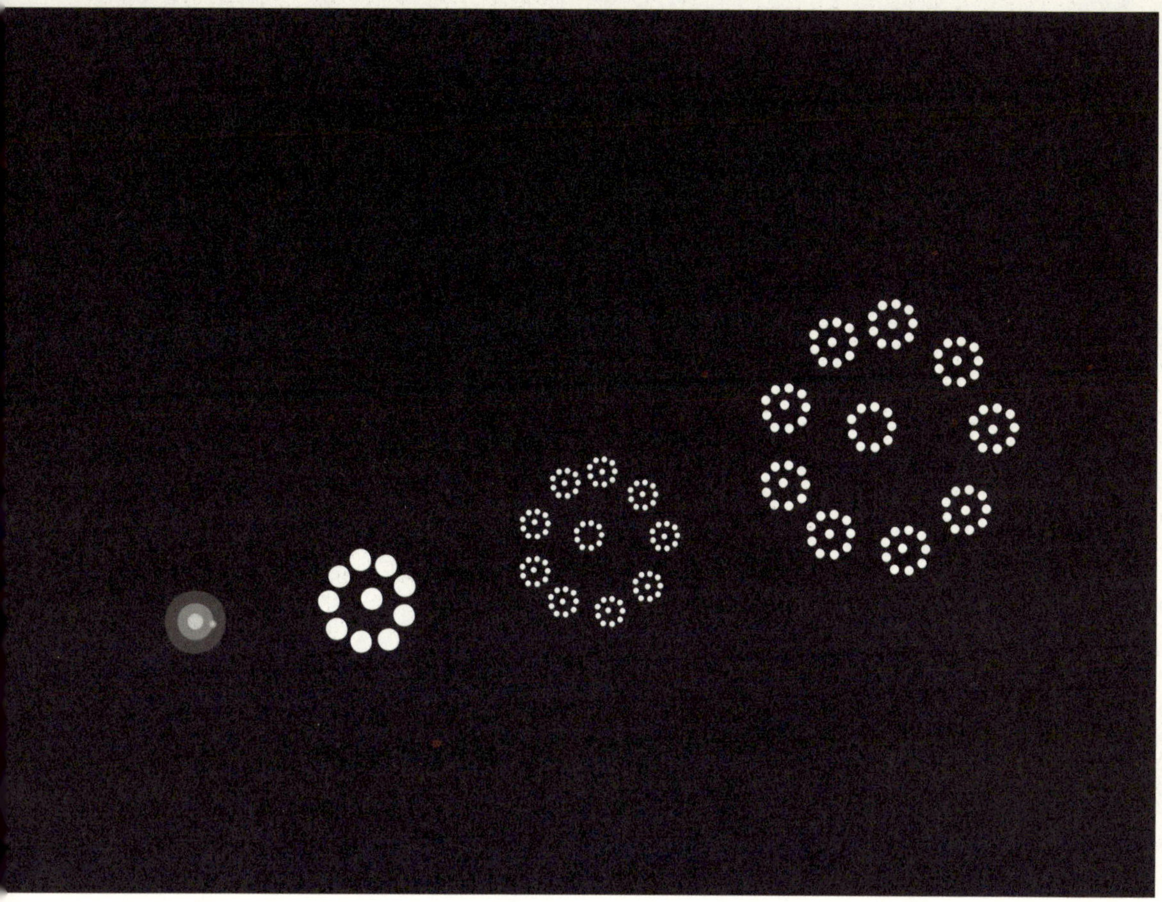

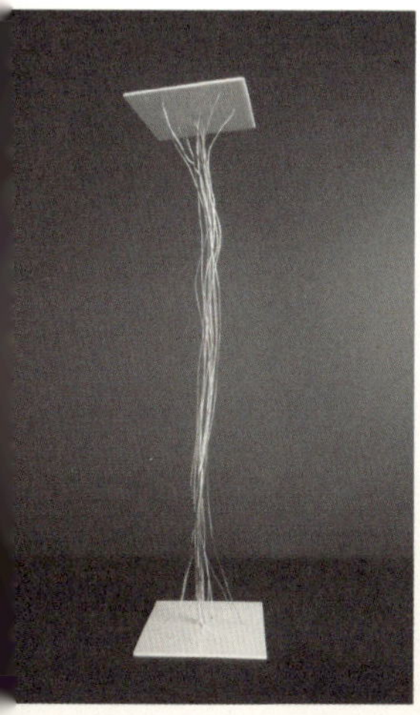

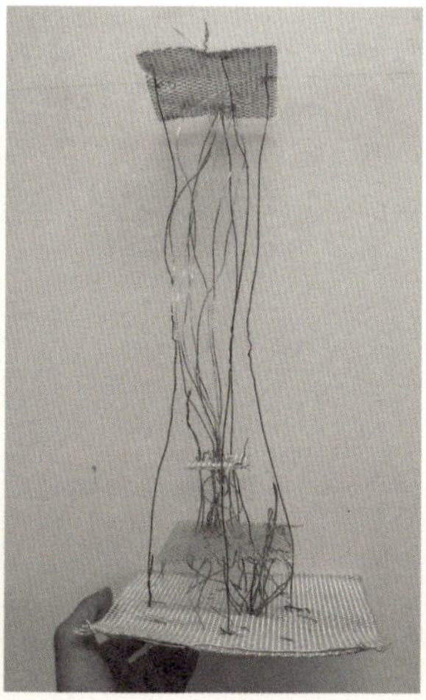

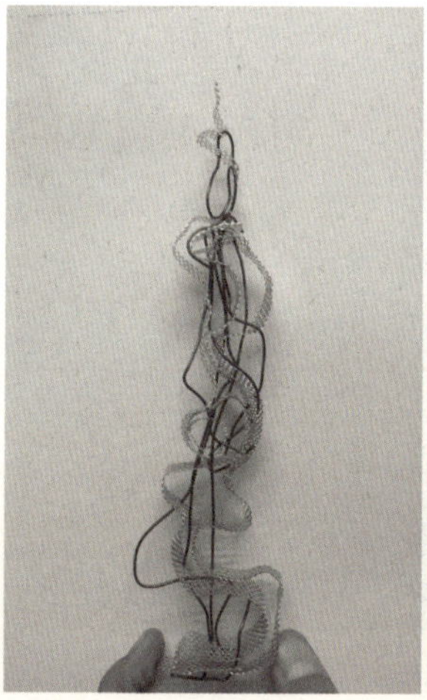

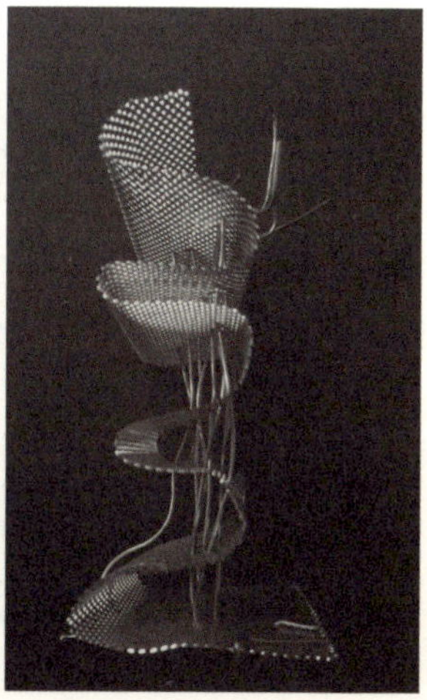

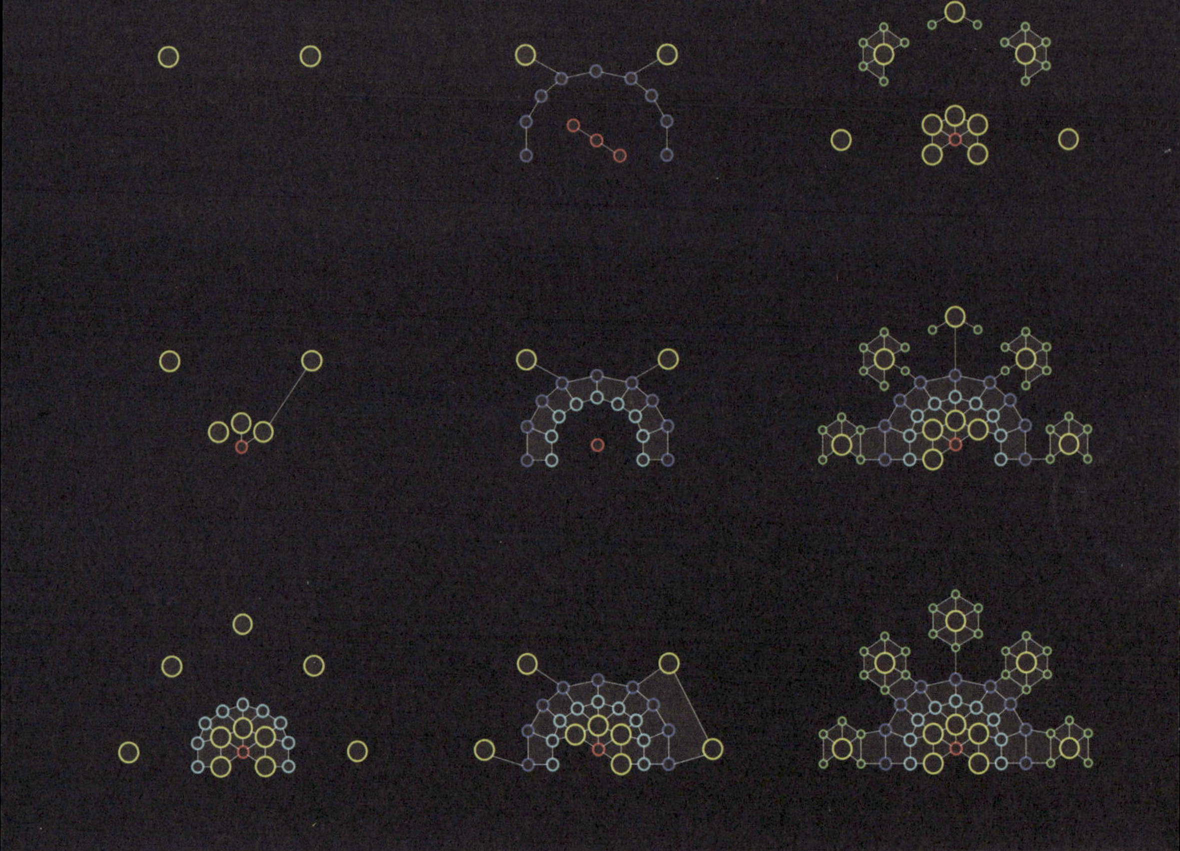

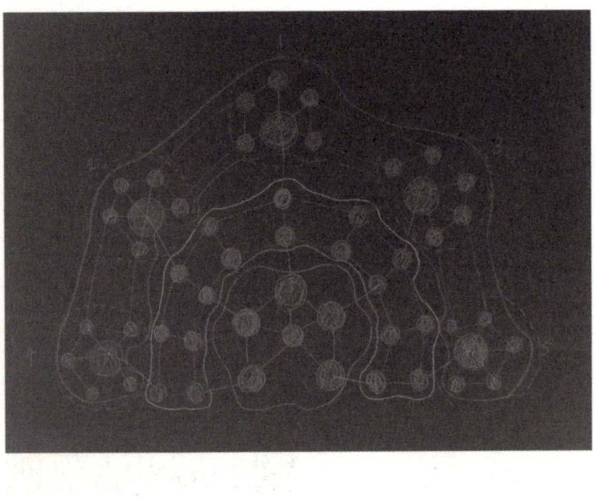

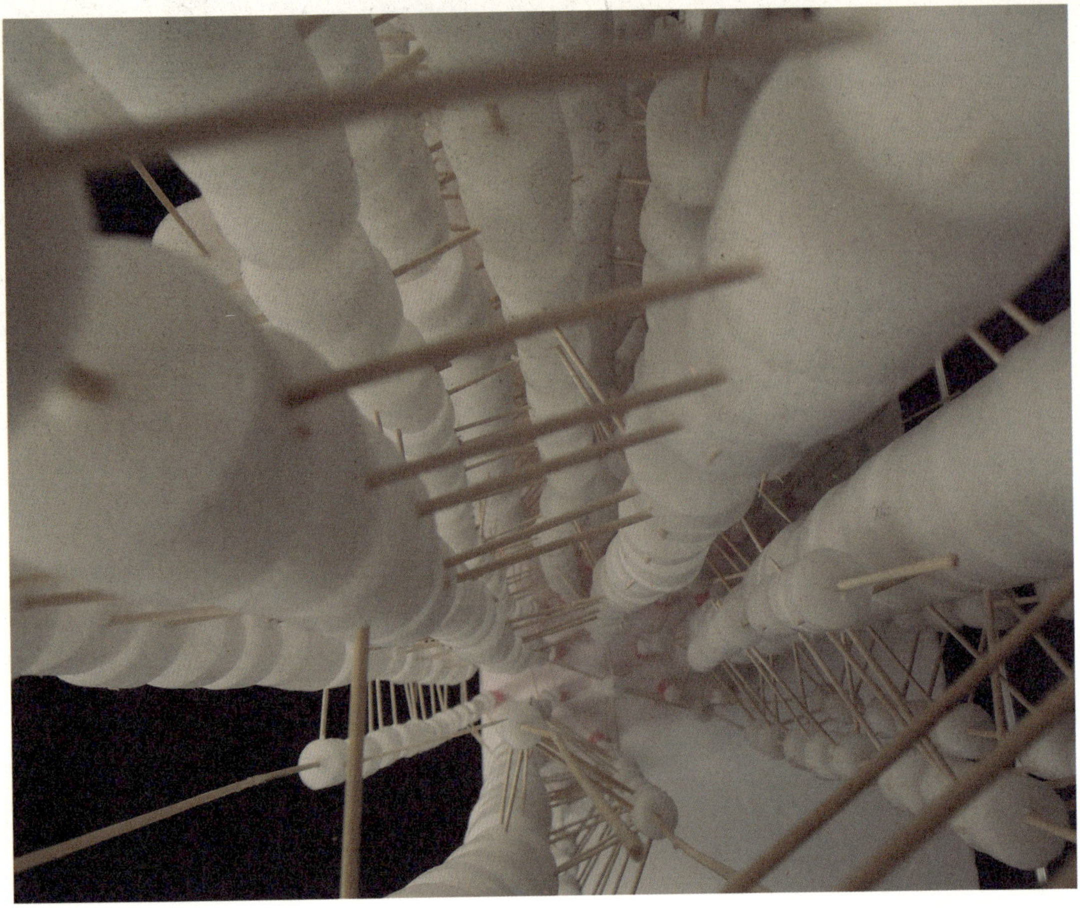

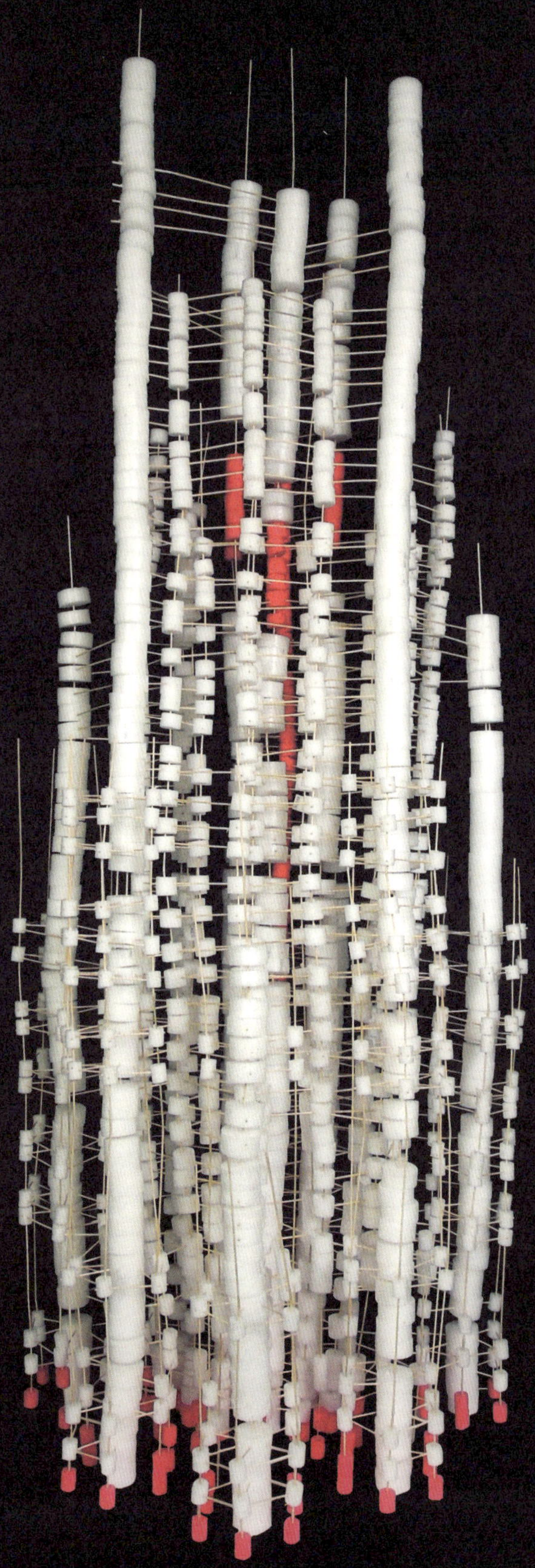

Sky High Zoo

Choi Han Sol
KOREA

Yongsan has been an inportant center of transportation, sovarious programs and event places are placed between complex flows of traffics, like cars, trains, subways, and pedestrians. To control this kind of situation, I borrowed the way of making the 'WRINKLEs' with clothing. There are several points where the different flows meet and the density is increased, so I called them 'NODEs', comparing with wrinkles. And following the state of places near the site, I could put some nodes on the site. These nodes made the density and tension on there. On the nodes, I also applyed the rule of wrinkle vertically. With tension and compression, I tried to create vertically complex and huge space, and then place the essential, but ordinary programs like office and residence, outside of that space.

The 'ZOO', inner space is some kind of vertical topography,and vertical public landscape, and there are some cores and bridges which maintain this highrise space. Inside the zoo, public programs like librarys or schools will be dangled to the structures.
Hard-boiled urban order and soft rural fabric, the strict cores for urban high speed and curved paths for rural slow walking, all these different conditions will be mixed up in this vertical zoo.

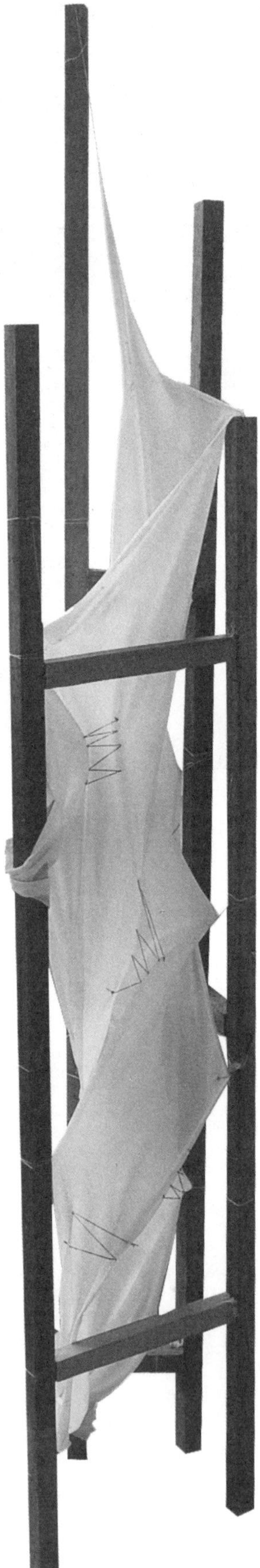

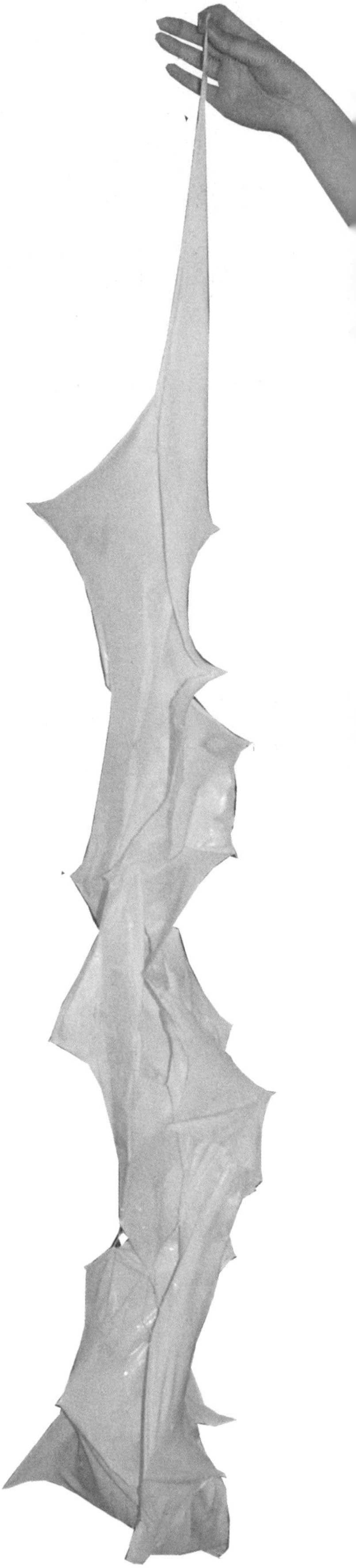

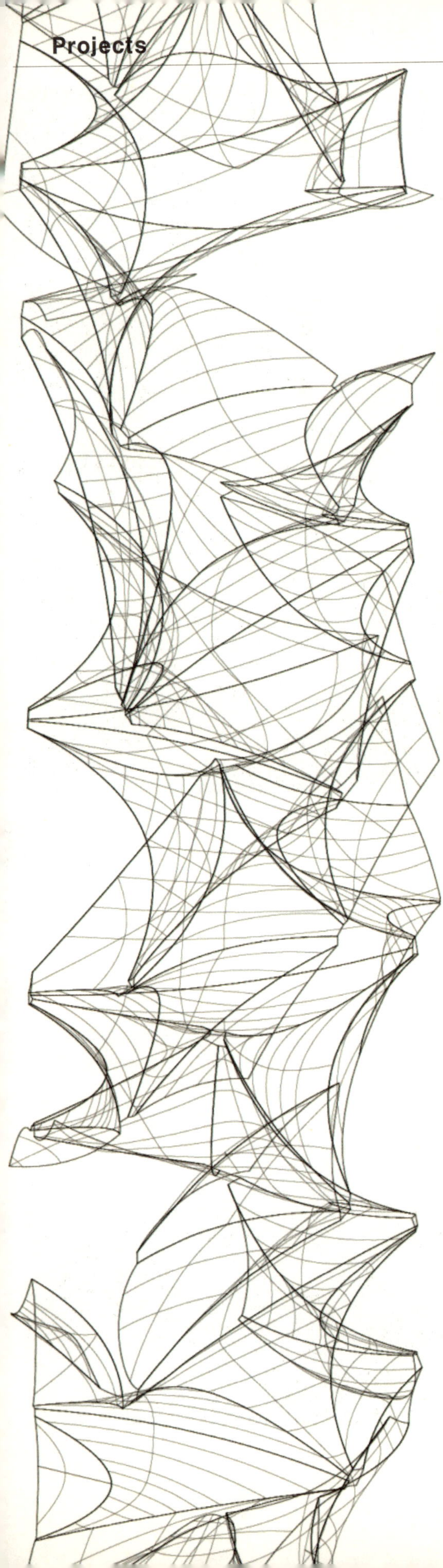

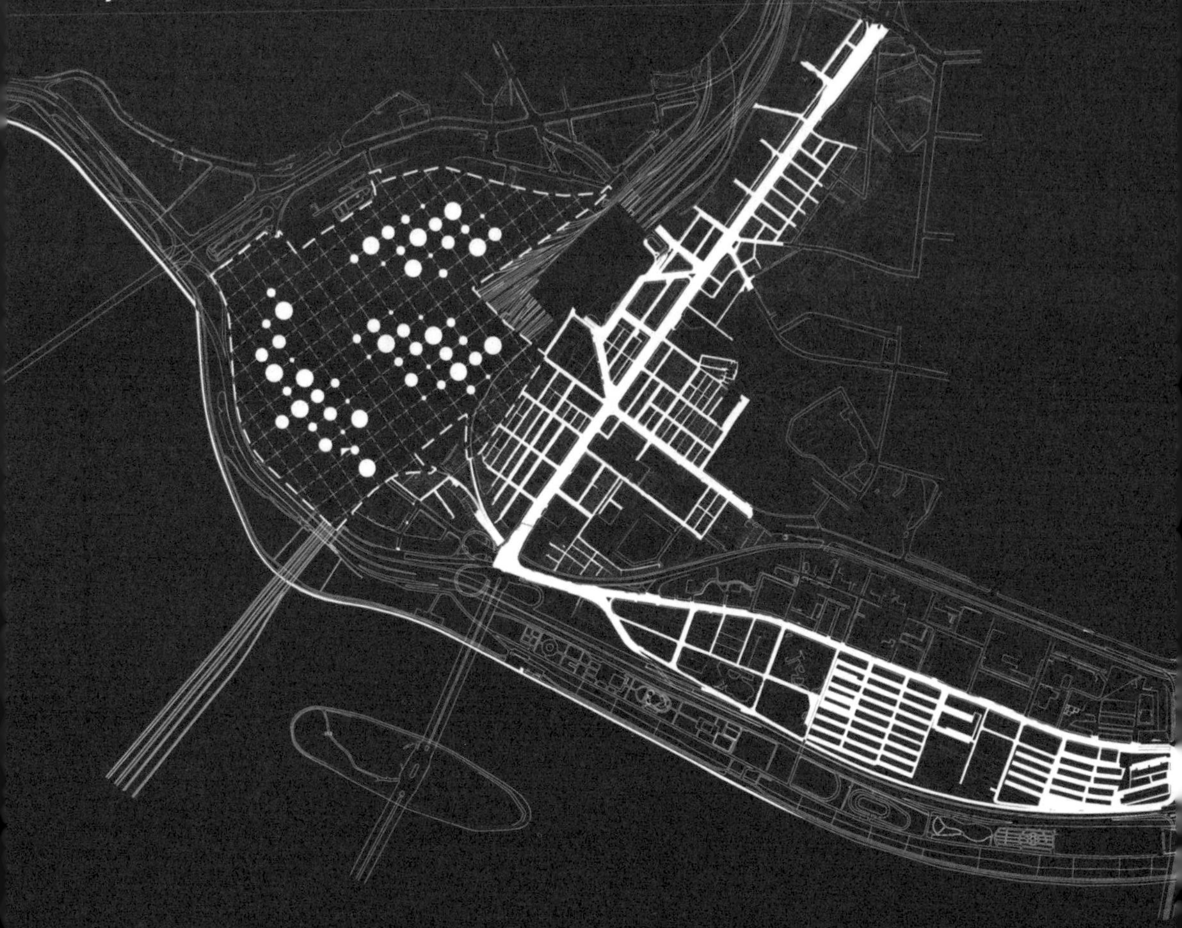

Soft Grids

Song Hye Jin
KOREA

While I found out earlier oil well in Arab. The oil well was put on the ground which had oil. When I reversed this diagram, I could get the clue about how I could make verticality.

At the research about history of Yong-san, I found out something about urbanism. That was grid.

In 1895, in Yong-san, there was nothing but sand. Yong-san was just port for people and material going to palace. At that time, this area was really important as passing road going to palace. And then, in 1910, Japan started to colonize Korea, they made railroad and tram line going to palace. As Japanese army took over this area, Japanese street was made and Yong-san station district was urbanized with grid. In 1965, after Korean war, US army took over this area, and then, Japanese street changed into foreigner street, after then government urbanized riverside area with housing on grid.

I thought urbanism could be started from grid so I decided to put the sixty meter grid on the site. Before putting program on this grid, I brought our master plan on the site, which was about "on" and "off". "on" means urbanism, "off" means rural. Riverside, nearby market, and station could be "on", so I started putting program on grid along this logic.

On 60m grid, I put four different size circles Because I want to find relationship between S, M, L, XL when they are gathered. And then, I tied outside circle and inside circle in different way. Circles will be cores which is infra, structure, housing, etc. tied outside circle would be slab which have public program such as garden, caf , playground, market. Tied inner circle would be slab which have program such as library, hospital, university.

Indubitably, it is megastructure. But the thing that I tried to do was another logic which could compose urban system. The logic that I've made in this project is quite mechanical and mathematical, but I think that this way could be another experimental attempt to compose vertical urbanism instead of typical tall building.

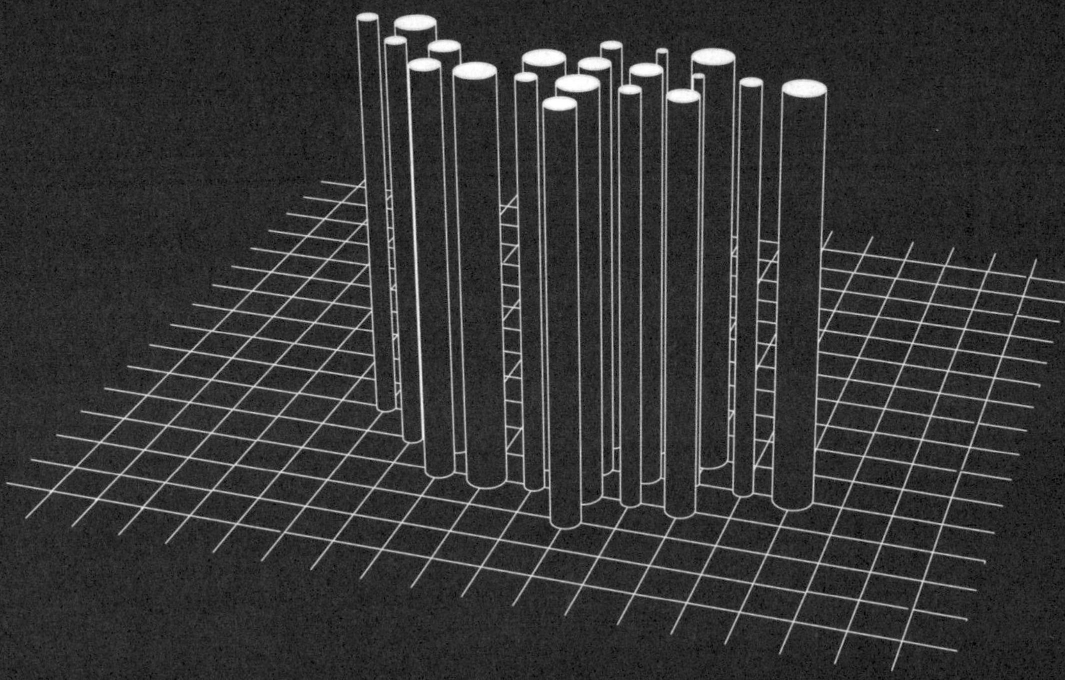

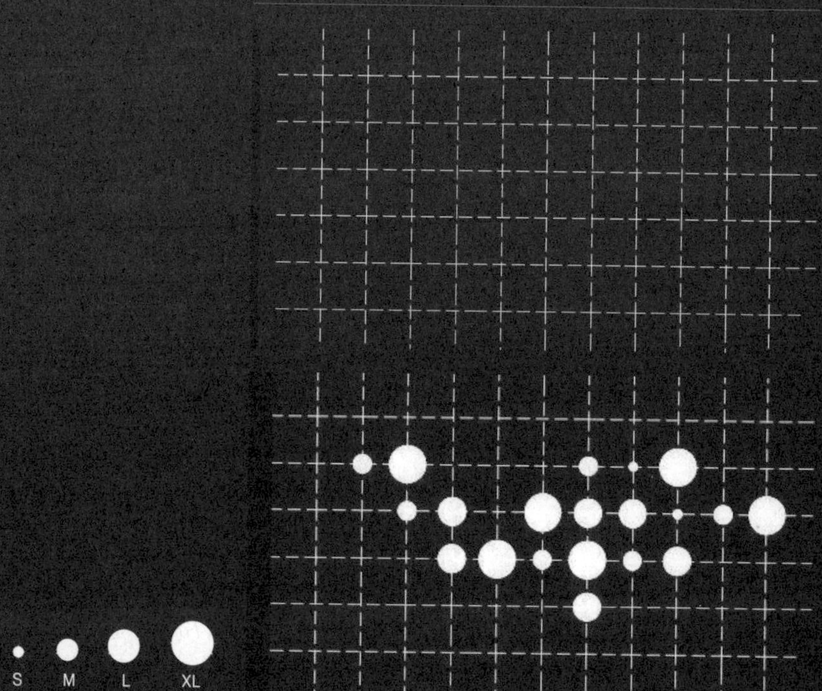
S
M
L
XL

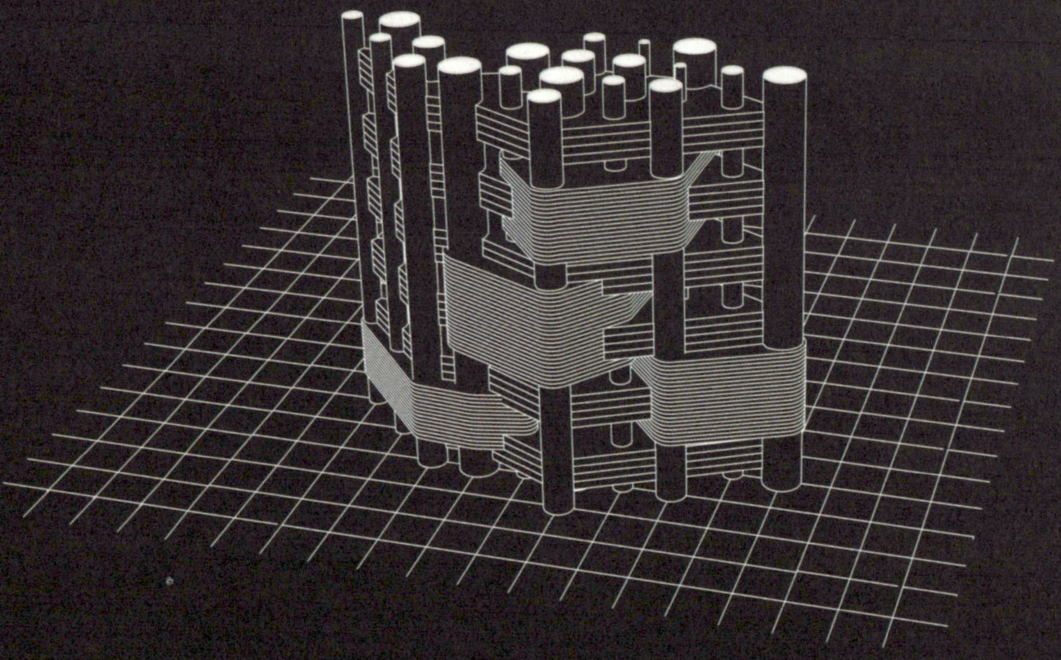

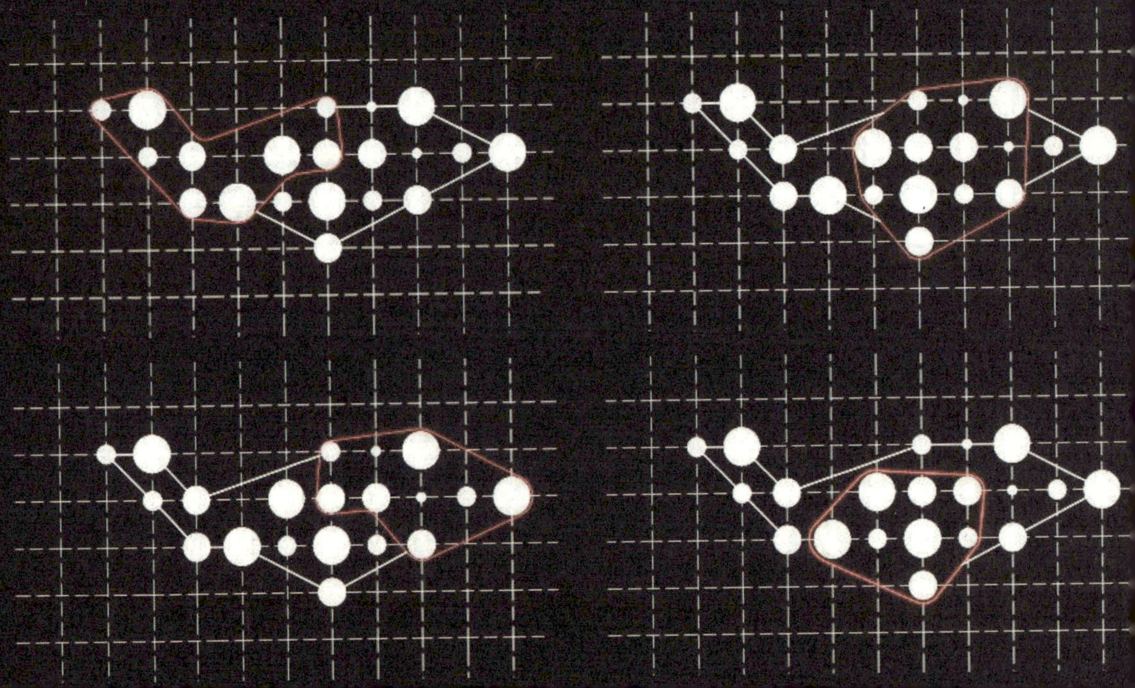

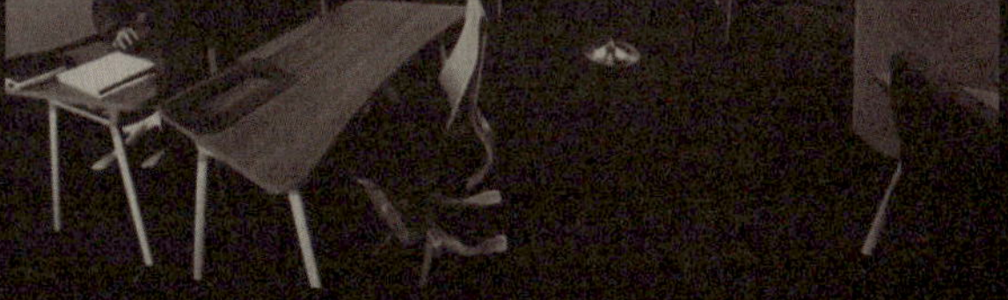

Floating Street

Lee Myoung Ki
KOREA

Project is started from Walker Evans's a picture (I named it as "connection" for 'Reference Book' we made previously). Consisted of silos and grain factories, the picture shows possibilities of connection of buildings in air.

It's not easy to handle whole area of YIBD. I tried to form a type of cluster. In present days, we can see the cluster of high rises, basically only thing they connected are lower floors of buildings. This situation makes upper part of high rise isolated.

Then I studied about connection of buildings in air, how they connected and what connection parts could be. Connection parts could be merely lifted up to make space below. And connection parts of building could form a loop and a street.

In section study, system of cluster becomes specific. First, 4~5 towers are connected in air, its connection parts are lifted up to make huge open space below. All of ground level commercials are located in underground area. It will release density of city, and provide huge ground level park. Second, connection in air could be a simple solution of isolated high rise's upper parts.

Connection parts will be second access to the buildings. And they could be used as a public space like a market street, a tour site, gallery and library. And upper side of connection provides an artificial landscape. Third, tower will be more stable than just stood by itself. 4~5 towers can form a shape of a table or a chair and it makes them structurally stable.

In tower's shape, to provide more ground level open spaces, foot of building is minimized and top of tower also minimized to give more light to the artificial landscape. And they are tilted to form a chair shape. And connection parts are maximized make shape of tower belly.

With the system, I tried to put clusters in the site. The site was railroad depots and they were treated as an unattractive spot. Existing buildings of surrounding site show their back to the site forming some kind of wall. This situation makes the site an island. To break this situation clusters are placed along the boundary, extending arms to surrounding areas. Especially, North-west part of the site is residential area. Even they are close to the site, they have a little relationship with the site. Cluster in this place have a tower on the residential area. And promote accessibility between two areas.

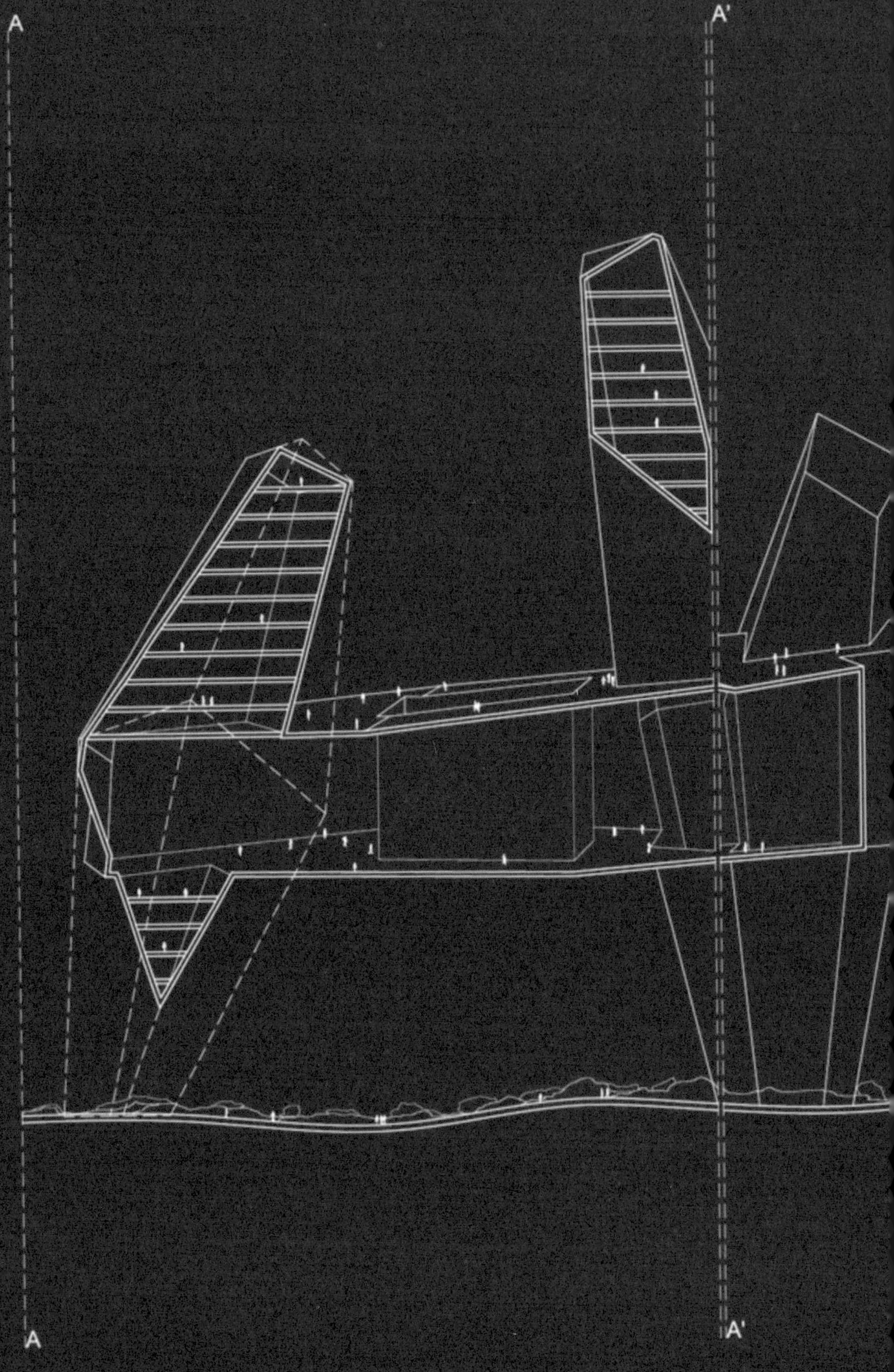
A
A'
A
A'

B
B'
A
A'
B'
B
B
B'

Porous Brick

accurate accumulate

Baek Do Hyeon
KOREA

Project starts with thinking about verticality. I found out when things are stacked, they made verticality; and typical high-rise building's skin is always glass, so I want change that strange tradition.

After research about verticality, I analyze Daniel Libeskind's master plan, Where exactly same site.

My project begins. First master plan, and second choose one colony, design more detail.

At master plan session, I have several concepts, like burst building and create soft shadow, flip over the building and make free ground space, put public space at upper level so create gradation of public and private. So, start to make a 1:100 model and porous huge void emerge.

From that point, my project turn the way to create beautiful porous void. I make porous not by cut holes on mass, but by make units, stack in, cover in, then it create bigger units, and so on and on. So, I create multi-layered porous system.

24
Coca-Cola
24
375 mL
CAN PACK
Coca-Cola
real taste, uplifting refreshment

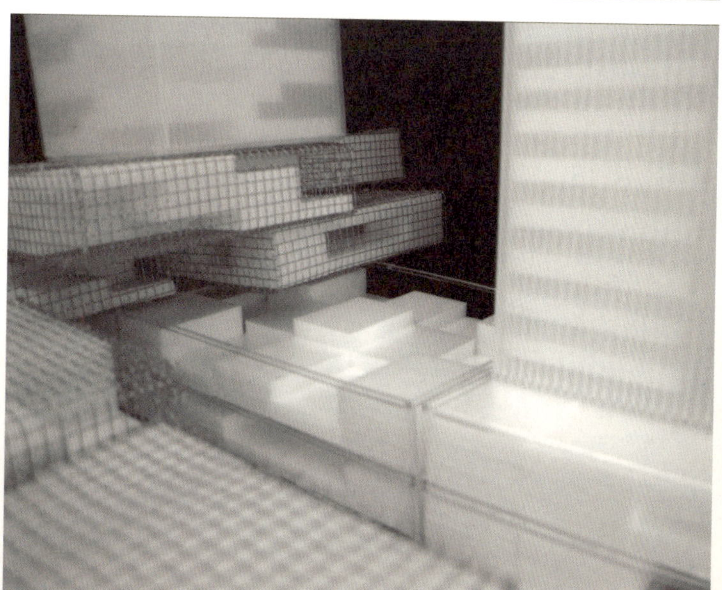

Physical Extraction

David Kriechmair
AUSTRALIA

The Project's conceptual idea is not the establishment of a finished architectural project but the development of an essential idea throughout a process of transformation to the very first step of something called "architecture".

Following "Vertical Urbanism" it all starts with the topics very basic analysis out of a context and continues to a practical analysis and development of the project's specific site with finally a definition of its highly activated areas wherefrom the so found nodes of extremely high activity are extracted.

Based on this the node is taken out of its context and throughout trial and error vertically developed just as a drop of ink in the water, which expands its whole volume from one point of extreme concentration: The node, which there from becomes the principle of a vertical structure and develops into architecture by its continuing analysis and manipulation.

Therefore at first a picture of the process is reduced to its' formal attributes and graphically structured by horizontal layers wherefrom the picture's formal properties can be translated into a three-dimensional structure which, by cutting and extracting, is brought into solid model and physical shape by adding its' vertical layers. Indeed the transformation from virtuality to reality reveals its structural debilities which there from make necessary the introduction of a stabilization by inserting vertical cores the process' rerun, at which's end a by process generated shape finally is converted into a solid model and architecture in its purest implementation.

Just throughout this process of analysis and transformation the so generated final model reveals its very own roperties which from there on offer the base for the development of a more specific architectural concept in association to its specific site.

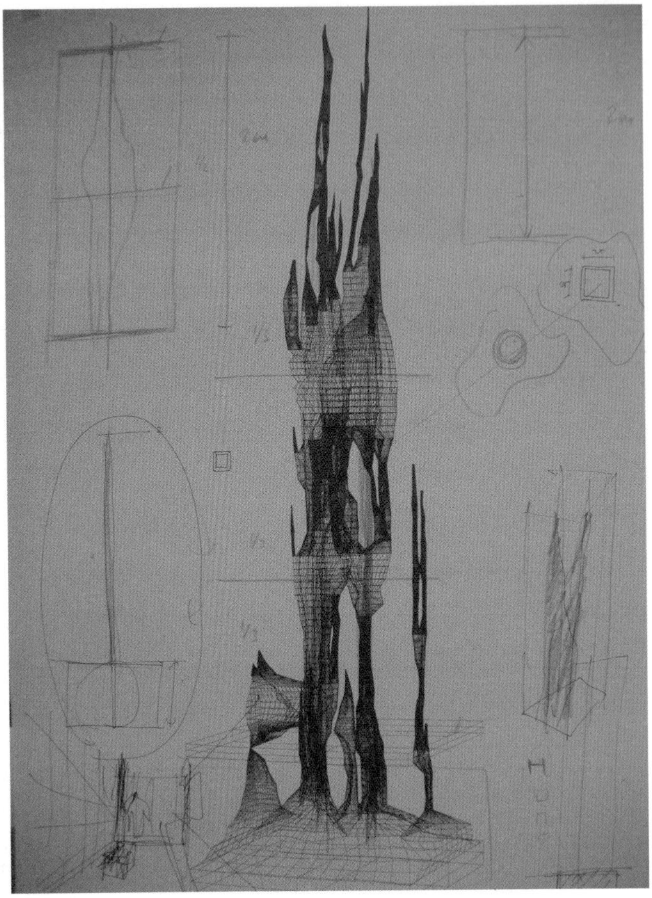

Eco-Urban Mutation : "A machine for living"

Lucia Espinosa De Los Monteros
ESPANA

[MACRO]

The first steps I did to approach the project was the Macro analysis research of the site from the perspective of how the movement of the people around the world is related to Yongsan site in Seoul. The research was based on five elements: the analysis of the relationship between the high grounds of Seoul; the national network connections by road, boat or flights; the world related to the site following four parameters (cultural, business, tourism, commercial); the flight connections between Seoul and the rest of the world; and the relationship between the high ground of the cities and the latitude, related to the different climates and altitudes.

[FUNCTION]

The macro analysis ended with some movement conclusions. Searching for different ways of vertical movement for a living city I found that the cell mutation was the perfect way to approach the analysis of a vertical moving system. The Amoeba cell mutates as a function, it moves searching for it needs. Seoul is a developing city which changes as time passes by, depending on what the people needs. The way of changing the shape of the city is by addition or subtraction of elements in a static movement. It was interesting to search for a new concept of living a developing city that changes with it needs, a mutant city.

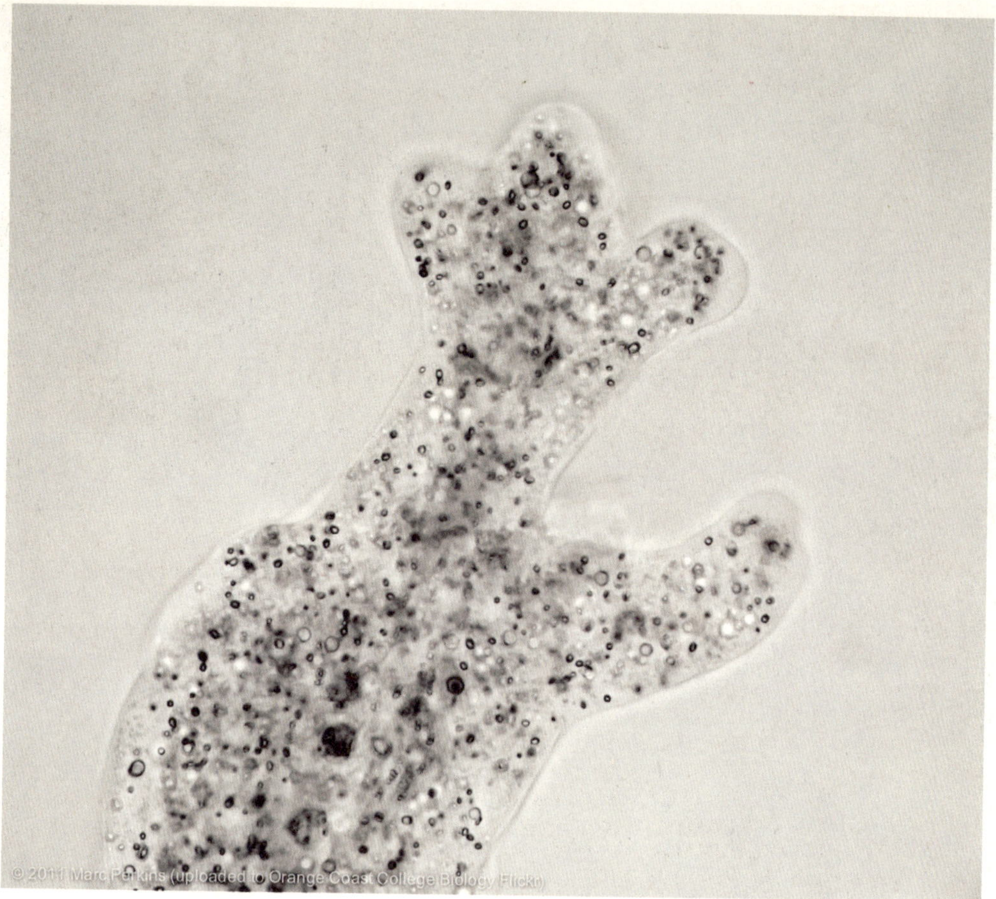
© 2011 Marc Perkins (uploaded to Orange Coast College Biology Flickr)

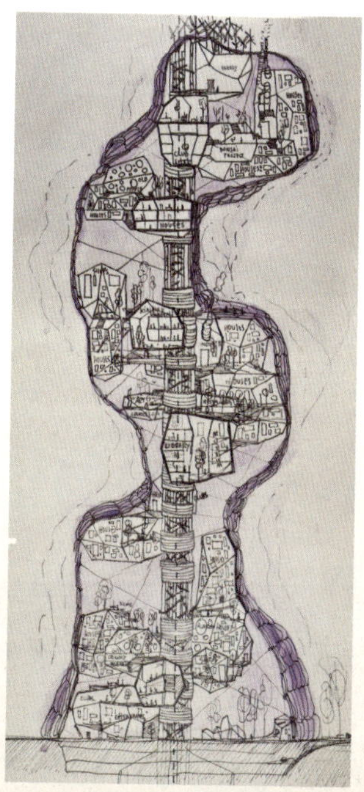

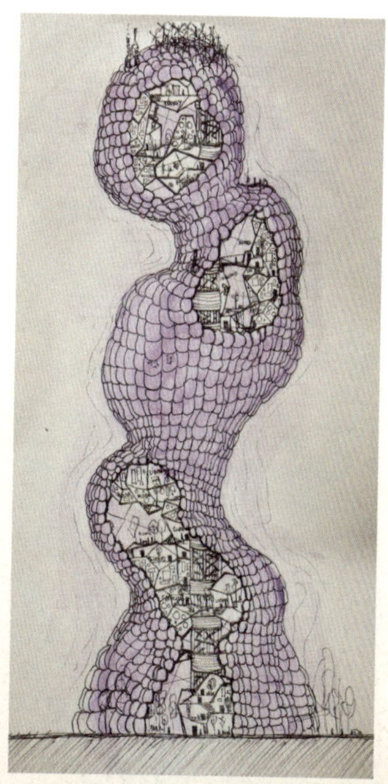

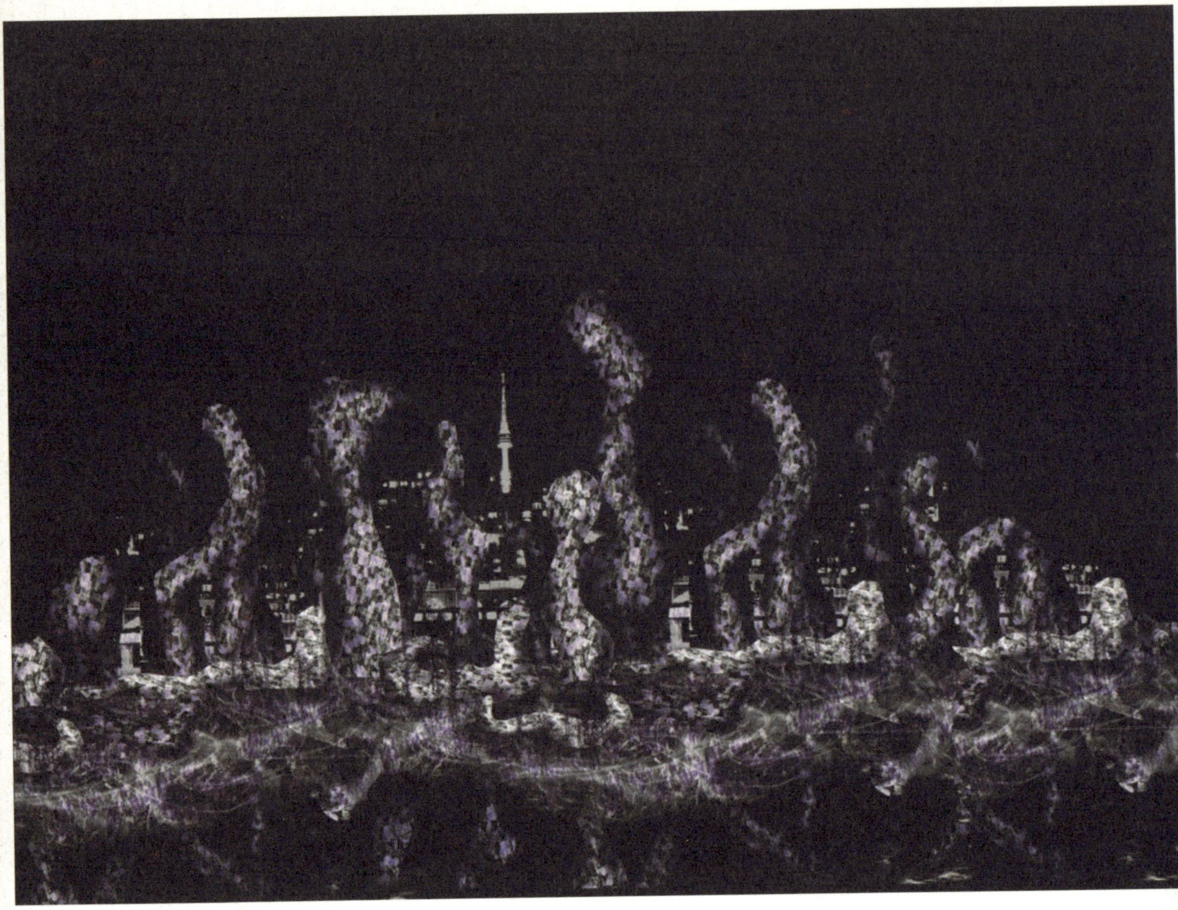

B
B'
C

B
B′

(PROGRAM&STRUCTURE)

The Amoeba needs three main elements to mutate: the nucleus which generate the energy and is the center of the movement; some protocells that moves around the nucleus searching for the food; and a membrane skin that changes it shape when the protocells move. The plan was to propose a new concept of living a city. Three thousand people are living on the vertical Eco-Urban skyscraper. The main idea of the proposal was to create a new vertical street where all these people could live with just few needs outside of the building. The structure is based on one main structure of columns with vertical connections, elevators, installations and stairs between them which serves as stabilizer; some public and private capsules moving around the structure; some green outdoors spaces; and a membrane with a snakeskin structure. The program is what people need, the function. And what people need changes. So the final purpose was to fix the program to the function. The function is to cover the three needs: housing for three thousand people (private use); indoor public uses as kindergarten, cinema, etc.; outdoor public areas as picnic, street market, parks··· which are generated using the space between the capsules; and a skin that covers all these elements that create an atmosphere around the private uses, indoor public spaces, and outdoor public green areas, but inside the building: a vertical mutant street.

(ATMOSPHERE AND MOVEMENT)

The snakeskin is one of the most important elements for the atmosphere of the vertical street because it has three different materials: opaque material, semi translucent metallic net with vegetation and a translucent metallic net. These materials filter the natural light to create a scene of outdoor space in the building. The private use will be the houses for three thousand people with two types of views: outdoor view; and the public outdoor views of the indoor green spaces which are characterized by a filtrated natural light. The public indoor spaces will adapt for all the different ages and styles of life: hospital, old people residence, kindergarten, bonsai project, cooking lessons, cinema, aquarium, museum, library, shops, restaurants, coffee shops, night markets, clubs, pubs, industries, energy resources etc.. These capsules can change its shape, program and position during the days, months and seasons depending on the needs of its inhabitants. The outdoor public areas will have the natural filtered light with different recreational spaces as picnic, street market, parks, playgrounds, terraces, jogging area, concert stages ···
A machine made for living.

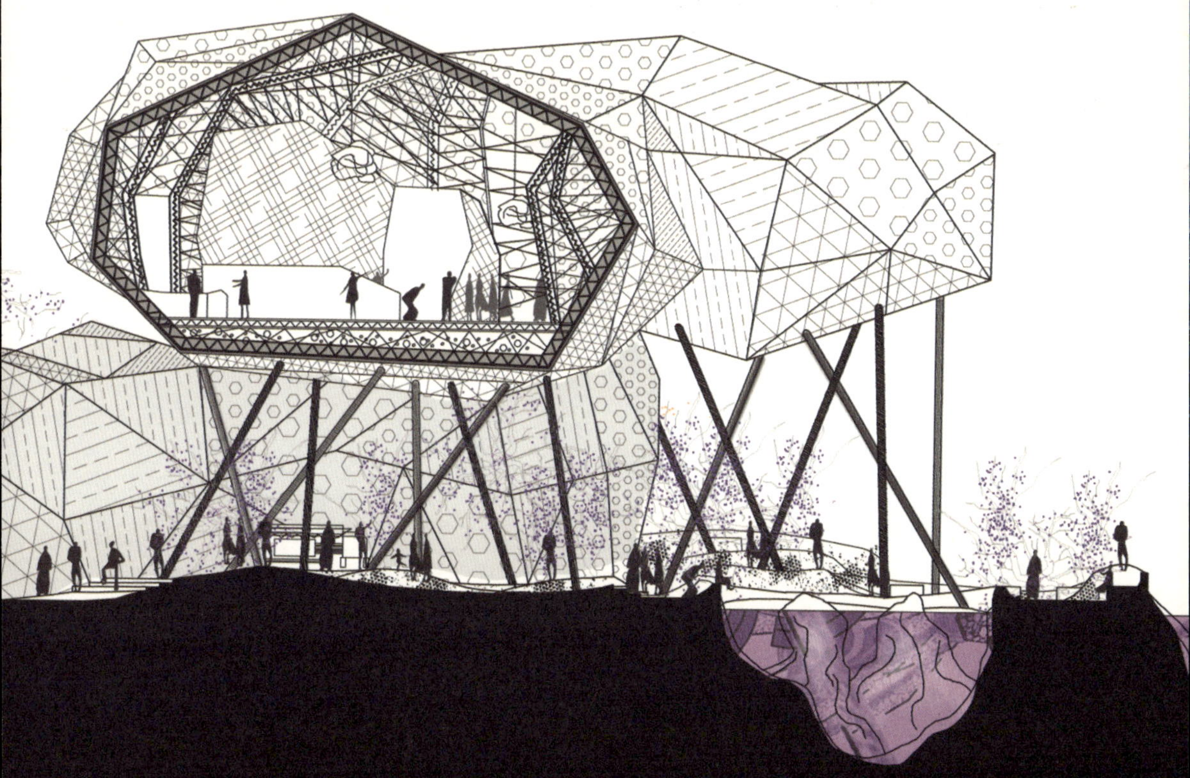

Inhabiting Structure

Marina Kindelan
ESPANA

The Project is focused on four main ideas, which I will call "rules". By combining these rules in order to solve different situations, a "system" is generated. It is this system what generates the project.

The four main concepts, or rules, are: 1. Setting free floor areas, 2. Circulations 3. Program reversion 4. Structure.

The main intention is to change the preestablished concept about what life in the city centre is like. As we know, as time passes cities are more invaded by buildings and green areas are lessened. So there is very little free space in the cities. This is the reason why the project is oriented to produce minimum impact on the floor. This intention's will is to free space up for leisure which is missing in nowadays cities. This is, in fact, the reason why people sometimes choose to live in the outskirts.

In addition to this, and related once again to the idea of freeing space up, in the project there is a separation between circulation areas from leisure ones, so that the latter are not interrupted by the former. Thus, in the project, space is divided into two areas: the passive and the active one.

As it was said before, the Project is double faced, so when program is distributed, each activity will have its place. Opposite to common cities, public activities are in the top part of the building. In this way we achieve making the building more active, while we leave some space free on the lower part. This means that a new circulation concept is settled: From upwards to downwards.

To finish, as a linking element there is the structure. This structure is the point of departure to generate all the urban area. It is not a simple structure, but it is functional: It serves as a transition area between public and private space. This structure is bond to create flexible spaces organically grown to give an answer to future necessities. This is what the project supports about structure and about facilities and future growth possibilities.

명성기업
초
LOEWS

옷 가게

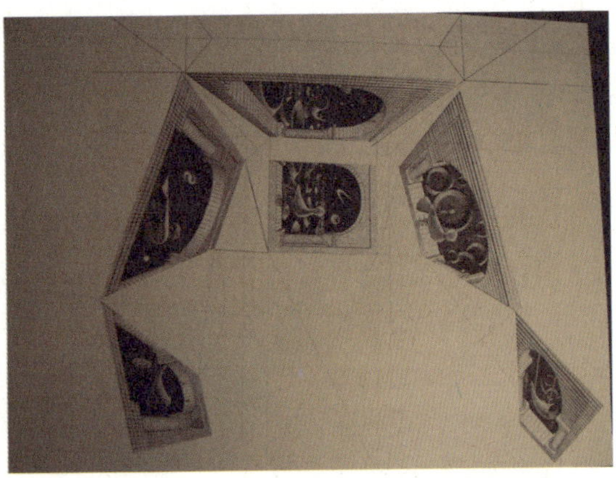

Folded decoupage

Ryu Hwan Chan

It starts with cartography and its dialectic with land perception to maps, urbanism and architecture through out the history.

From research the project stands on the metaphysical allegory of oragami where a plain, which is neither cut nor fragmented, is folded, as it keeps its whole and continuity, and build-up a 3-dimensional reality, verticality.

The map or drawing is, then, expands and reveal its 'structural area' where should be folded to build up a 3-dimensional object which it represents.
Such 'structural area' is narrative.
Narrative is not hidden.

It is everywhere.

Its omnipresence makes it subtle and hidden inside the 'structural area', when cut the metaphor one don't draw it on drawing since it is the matter of course; itself is the potential of making and reading of a drawing.

Through Oragami, verticality no more being a fragmented reality being stacked over a formal plain of reality.

With complex, elaborated folding lines of hidden section of plain, 'the structural area' the planar continuity keeps its whole to vertex.

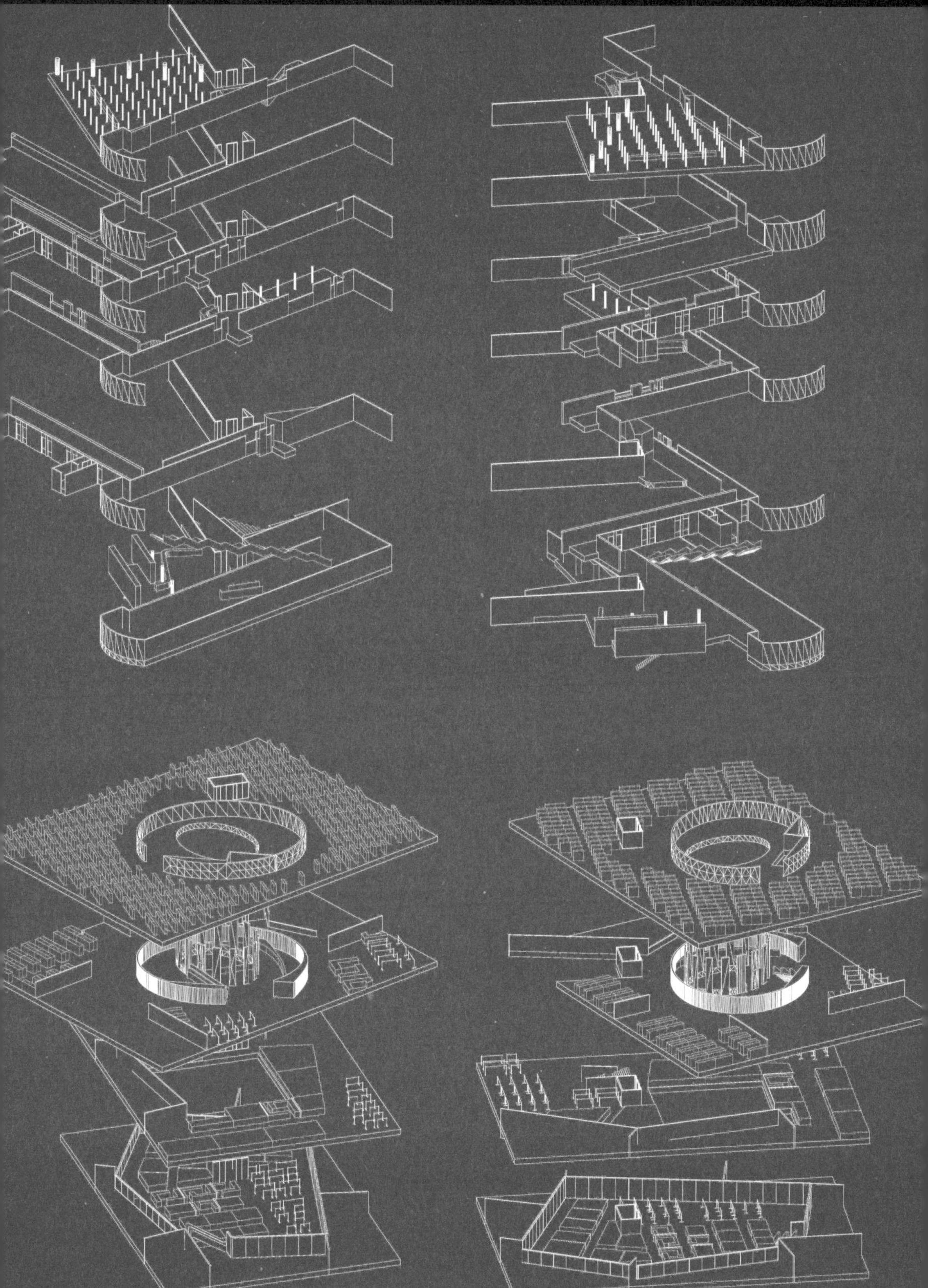

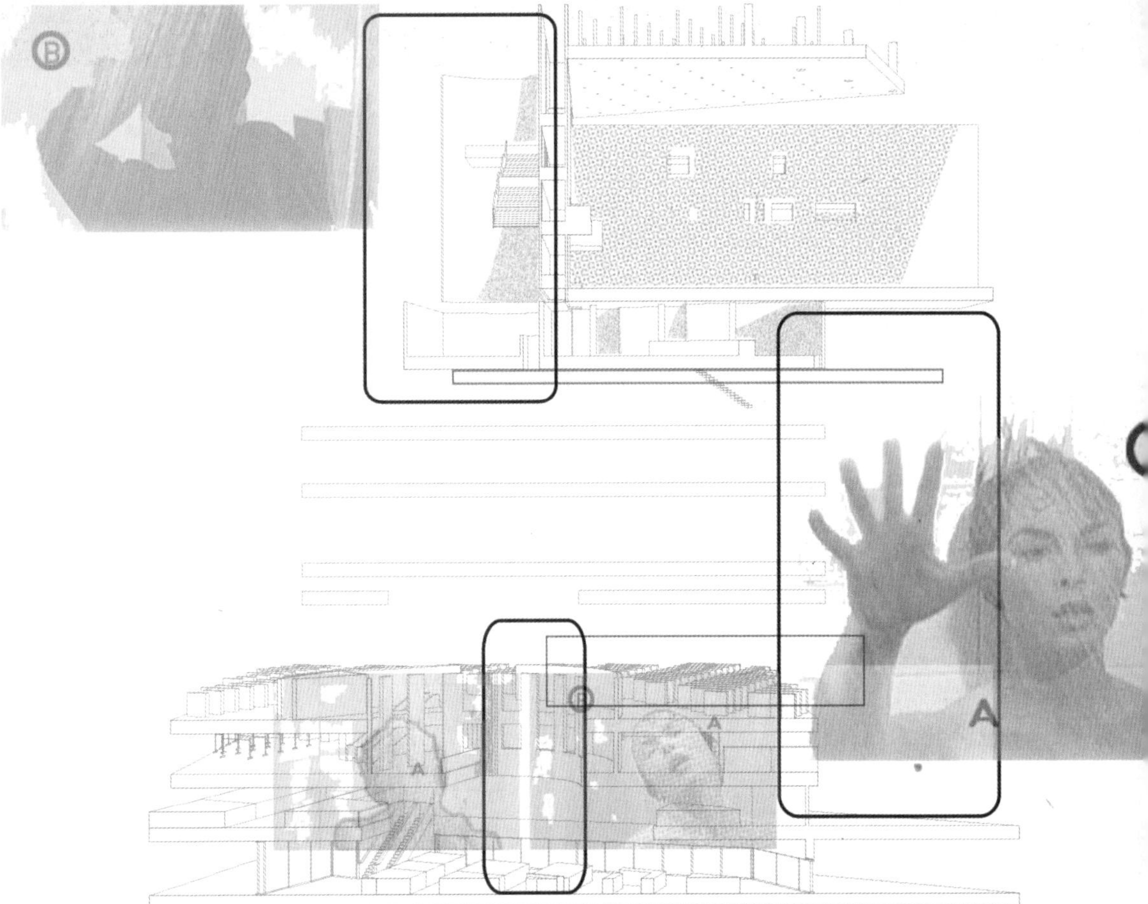
B
A
A
A

The End

Context

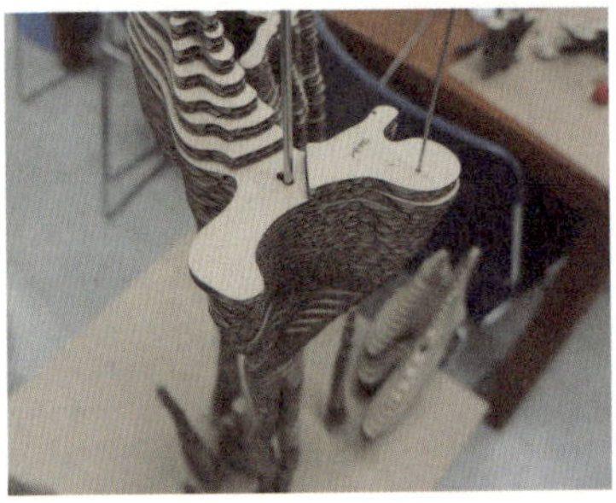

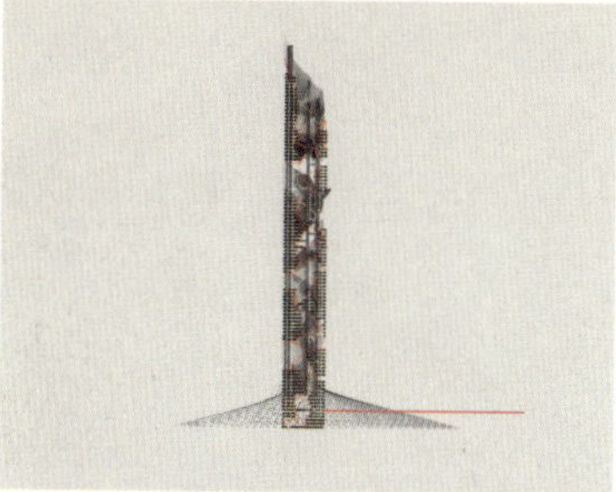

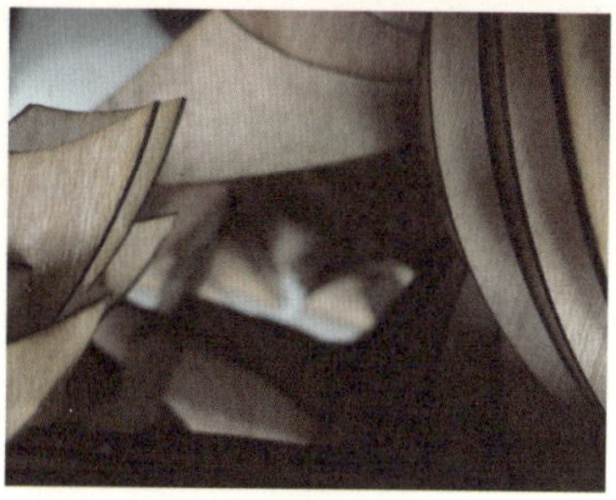

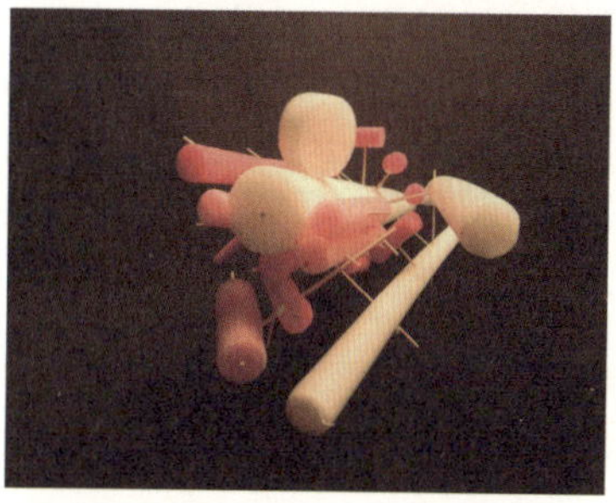

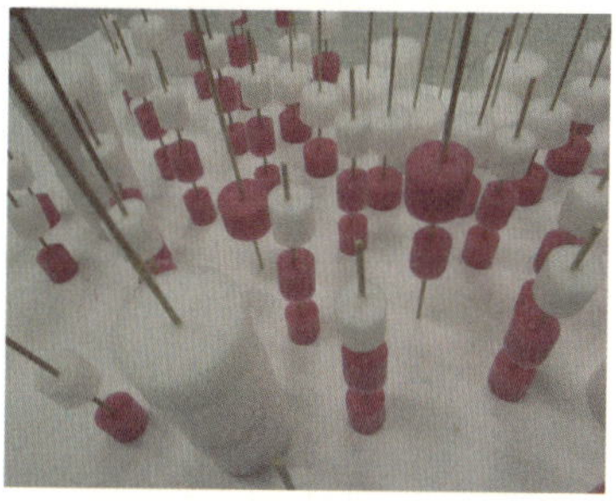

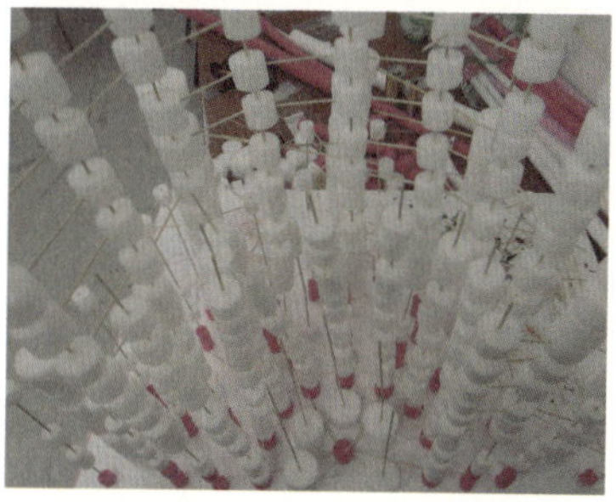

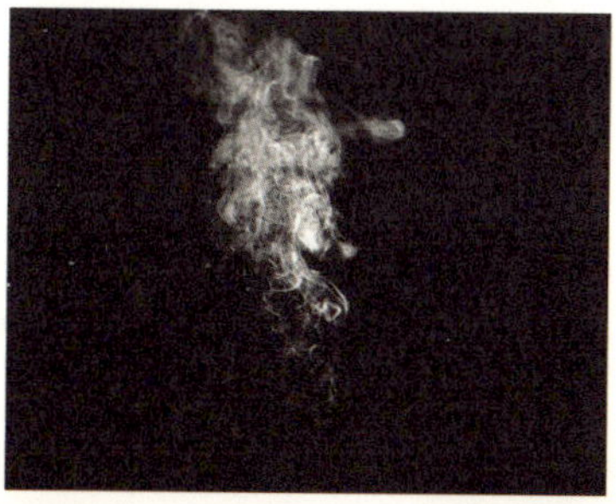

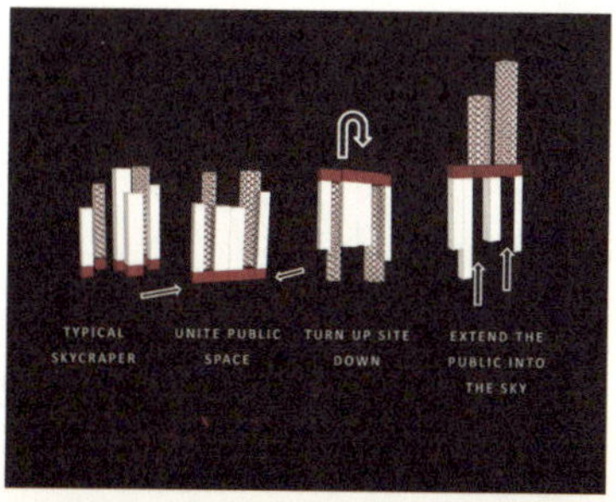
TYPICAL SKYCRAPER
UNITE PUBLIC SPACE
TURN UP SITE DOWN
EXTEND THE PUBLIC INTO THE SKY

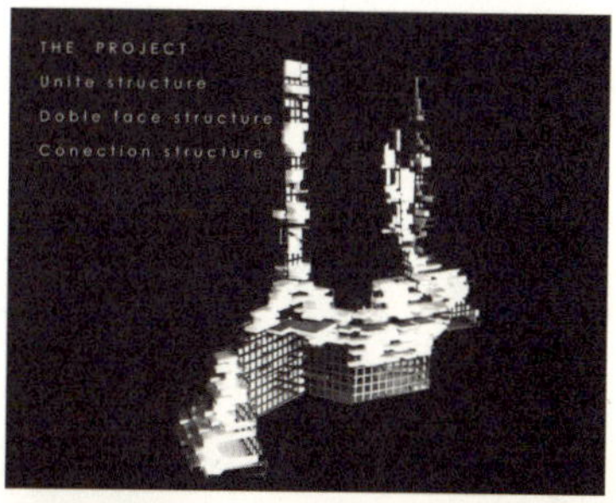
THE PROJECT
Unite structure
Doble face structure
Conection structure

BY MARINA KINDELÁN

STATIC
DYNAMIC

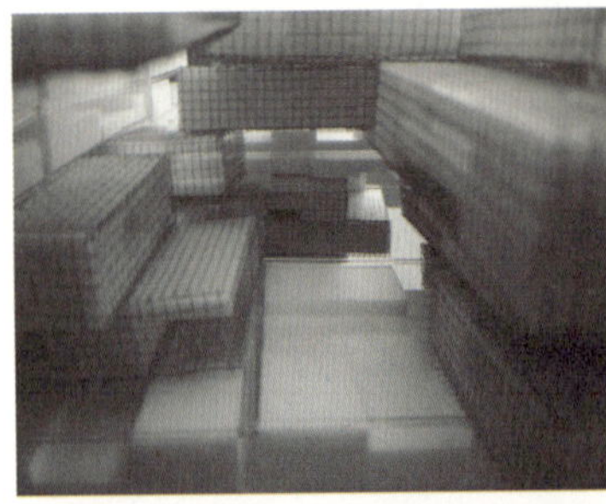

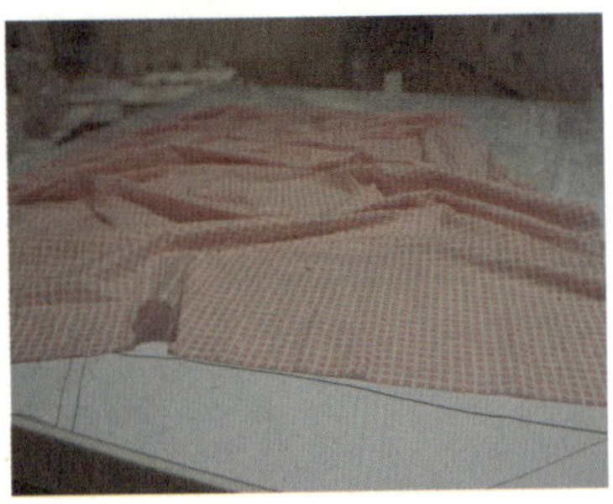

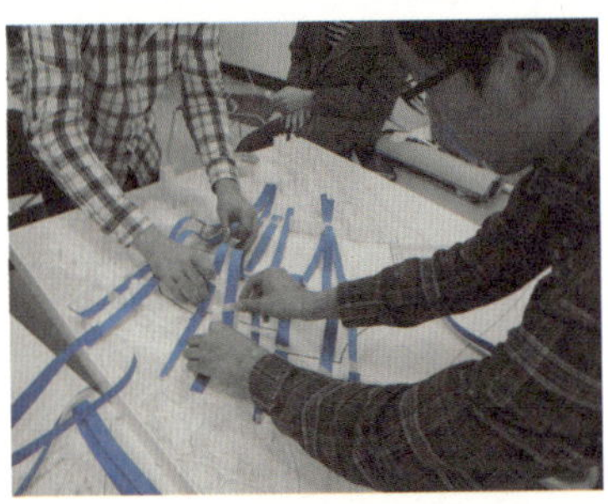
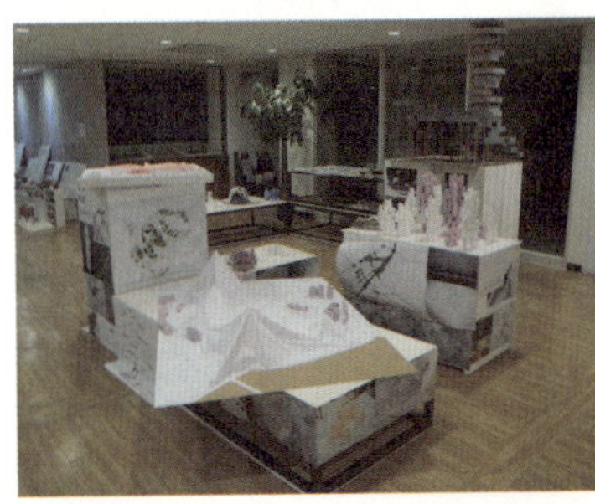

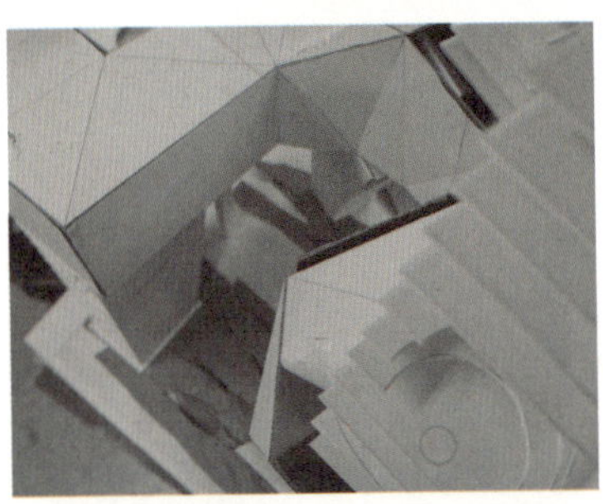

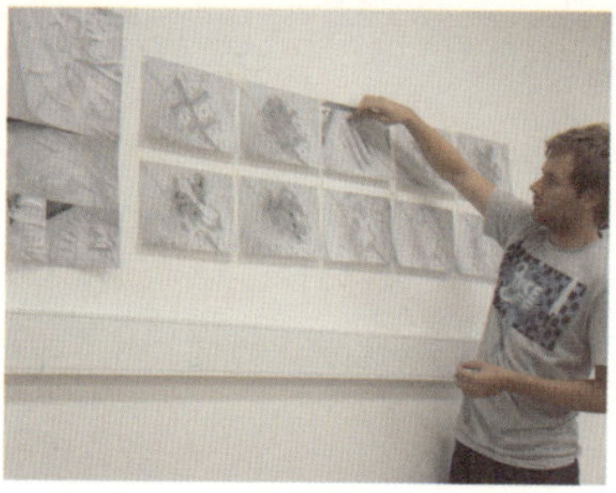

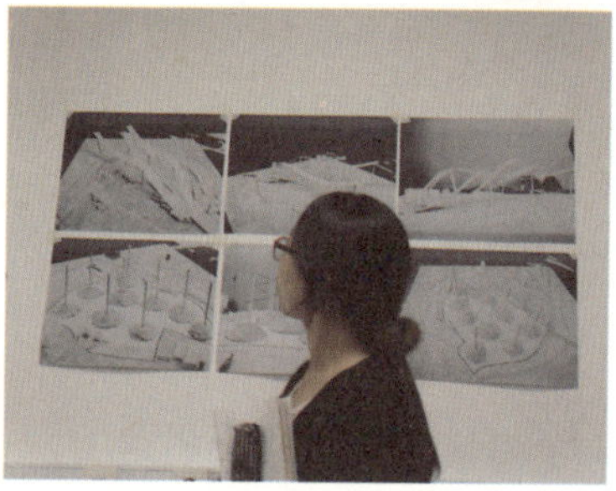
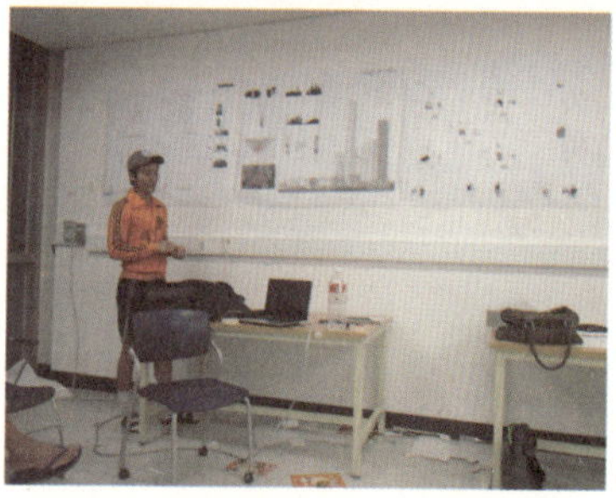

VERTICAL URBANISM

Author Peter W. Ferretto
Assistant professor
Department of Architecture
Seoul National University
Korea

Contact 1 Gwanak-ro, Gwannak-gu, Seoul
151-742, Korea
T. +82 2 2880 7051
F. +82 2 2871 5518
E. ferrretto@snu.ac.kr
W. www.pwferretto.com

Designer Lee Minho
Masters course
Department of Architecture,
Seoul National University
Korea

Printing Macprocess
T. +82 2 2278 2350

Provider Spacetime
www.spacetime.co.kr

ISBN : 89-5592-232-9

저자 피터 페레토
서울대학교 공과대학 건축학과 조교수

펴낸곳 151-742,서울특별시 관악구 관악로 1

전화. 02 2880 7051
팩스. 02 2871 5518
이메일. ferrretto@snu.ac.kr
홈페이지. www.pwferretto.com

디자이너 이민호
서울대학교 공과대학 건축학과 석사과정

인쇄 맥프로세스
전화. 02 2278 2350

공급처 시공문화
www.spacetime.co.kr